T0345138

Sources
in the History of Mathematics and
Physical Sciences

4

Sources
in the History of Mathematics and Physical Sciences

Editor: G.J. Toomer

VOLUME 1

Diocles on Burning Mirrors
The Arabic Translation of the Lost Greek Original
Edited, with English Translation and Commentary by G.J. Toomer
1976. ix, 249 pages. With 37 Figures and 24 Plates.
ISBN 0-387-07478-3

VOLUME 2

Wolfgang Pauli
Scientific Correspondence with Bohr, Einstein, Heisenberg, A.O.
Volume I: 1919–1929
Edited by A. Hermann, K.v. Meyenn, V.F. Weisskopf
1979. xlvii, 577 pages.
ISBN 0-387-08962-4

VOLUME 3

The Arabic Text of Books IV to VII of Diophantus' *Arithmetica*
in the Translation of Qustā ibn Lūqā
By Jacques Sesiano
1982. vii, 502 pages. With 4 Figures.
ISBN 0-387-90690-8

VOLUME 4

Descartes on Polyhedra
A Study of the *De Solidorum Elementis*
By P.J. Federico
1982. x, 144 pages. With 36 Figures.
ISBN 0-387-90760-2

P.J. Federico

Descartes on Polyhedra
A Study of the *De Solidorum Elementis*

With 36 Figures

Springer-Verlag
New York   Heidelberg   Berlin

AMS Subject Classifications (1980): 01A45, 52-03, 52A25

Library of Congress Cataloging in Publication Data
Federico, P. J. (Pasquale Joseph), 1902–1982.
    Descartes on polyhedra.
    (Sources in the history of mathematics and
physical sciences; 4)
    Bibliography: p.
    Includes index.
    1. Descartes, René, 1596–1650. De solidorum
elementis. 2. Polyhedra. I. Title. II. Series.
QA491.D473F43   1982   516'.15   82-19239

Typeset by Publication Services, Urbana, Illinois.
Printed and bound by R.R. Donnelley & Sons, Harrisonburg, Virginia.
Printed in the United States of America.

9 8 7 6 5 4 3 2 1

ISBN 0-387-90760-2 Springer-Verlag New York Heidelberg Berlin
ISBN 3-540-90760-2 Springer-Verlag Berlin Heidelberg New York

Preface

The present essay stems from a history of polyhedra from 1750 to 1866 written several years ago (as part of a more general work, not published). So many contradictory statements regarding a Descartes manuscript and Euler, by various mathematicians and historians of mathematics, were encountered that it was decided to write a separate study of the relevant part of the Descartes manuscript on polyhedra. The contemplated short paper grew in size, as only a detailed treatment could be of any value. After it was completed it became evident that the entire manuscript should be treated and the work grew some more. The result presented here is, I hope, a complete, accurate, and fair treatment of the entire manuscript. While some views and conclusions are expressed, this is only done with the facts before the reader, who may draw his or her own conclusions.

I would like to express my appreciation to Professors H.S.M. Coxeter, Branko Grünbaum, Morris Kline, and Dr. Heinz-Jürgen Hess for reading the manuscript and for their encouragement and suggestions. I am especially indebted to Dr. Hess, of the Leibniz-Archiv, for his assistance in connection with the manuscript.

I have been greatly helped in preparing the translation of the manuscript by the collaboration of a Latin scholar, Mr. Alfredo DeBarbieri.

The aid of librarians is indispensable, and I am indebted to a number of them, in this country and abroad, for locating material and supplying copies. But the chief source of material has been the conveniently located Library of the U.S. Naval Observatory with its remarkable collection of mathematical and other scientific periodicals and rare books, extending back to the 17th century, which I have been privileged to use. The librarian, Mrs. Brenda Corbin, has been extraordinarily helpful and many times came to my rescue in locating material and obtaining copies.

December, 1980

P.J. Federico
Washington, D.C.

My husband had finished this book and it was in the hands of the publisher when he died January 2, 1982. My family and I wish to thank Mr. Walter Kaufmann-Buehler for arranging for its publication, Professor G.J. Toomer for his careful editing and criticism and both Professor Toomer and Mrs. Janet Sachs for the final preparation of the manuscript. I also wish to thank my daughter, Joan Federico Kraft, for her assistance.

Bianca M. Federico

Editor's Note

Since I had read two earlier drafts of this work, it naturally fell to me to undertake that final revision which the author's death had prevented him from carrying through. I have confined myself to rearranging and presenting the material in a more accessible form, revising the references, and correcting any typographical errors and slips that came to my attention. The plan of the work, the factual content, and the opinions expressed are all those of the author.

I am grateful to Mrs. Federico for her cooperation and help in providing access to her husband's materials. I thank Dr. Heinz-Jürgen Hess and the Niedersächsische Landesbibliothek for permission to reproduce the relevant pages of Leibniz's manuscript copy of Descartes' work. I acknowledge gratefully the generous permission of Mr. Ioannes Papadatos to reproduce Figs. 24 to 36 from his book on the Archimedean solids. I thank O. Neugebauer for drawing several of the other figures. I am especially grateful to Janet Sachs for her very able editorial assistance.

G.J. Toomer

Contents

Part One
The Manuscript

1 Introduction

This essay presents the text and translation, with comments, of a Latin work of Descartes which exists only in a copy, made by Leibniz, but not known until 1860. The manuscript treats two subjects and is notable in several respects, aside from being a work of Descartes.

The first part of the manuscript attempts a general treatment of polyhedra, a somewhat neglected subject which theretofore had been considered mainly if not solely by treatment of specific individual solids, primarily the regular and semiregular solids and a few others. As to content, it contains some valid general results, including a theorem of intrinsic significance in solid geometry which was not known to geometers before the Leibniz copy was published, and from which Euler's famous polyhedron theorem can be derived (but by the aid of concepts not existing until quite some time later). In fact, the manuscript also contains a formula which can be considered an analogue of Euler's theorem and from which the latter can be derived as a simple corollary, but again only by means of a later notion. However, it does not, as contended or asserted by some, disclose Euler's theorem itself, either verbally or by formula, nor is there any intimation in the manuscript that Descartes was, or could have been, aware of that theorem. These conclusions will be evident from the manuscript itself and the discussion.

While the work of Descartes was apparently the first attempt at a general treatment of polyhedra, it was unknown for nearly 200 years and the first published general treatment was that of Euler, who was unaware of the earlier work, in two papers of 1750 and 1751. Comparison of the different studies of the two famous mathematicians offers some interest from the standpoint of the psychology of mathematical discovery. The basic discovery of Descartes was arrived at by analogy with plane figures, whereas that of Euler came about by induction from the solids themselves.

The second part of the manuscript considers figurate numbers corresponding to the regular and semiregular polyhedra and introduces a class of figurate numbers unknown to the Greeks.

The date the original manuscript was written by Descartes is not known; it most likely was before 1637, the date of the *Méthode,* and a date of circa 1630 is suggested here. It shows the thinking of Descartes on several subjects and may very well represent discarded applications of his *Méthode*[1] in an endeavor to develop a suitable mathematical example to include with the *Méthode.*

Part One of the present essay will discuss the manuscript as such. In Part Two, Section 5 will give some geometrical background for Sections 6 and 7, which present the translation of the first part and comments. Section 8 reviews the Euler papers on polyhedra and Section 9 concludes with a comparison of Descartes and Euler, with a note reviewing statements of various authors on the same topic. In Part Three, Section 10 gives some Greek number theory background for Sections 11 and 12 which present the translation of the second part of the manuscript and comments.

2 History of the Manuscript

A few dates and events in the life of Descartes will be noted, before giving an account of the history of the manuscript, a description, and a discussion of its date.

Descartes was born March 31, 1596, at La Haye, near Tours, France. When he was eight years old he was sent to the Jesuit school at La Flèche. In 1612 he went to Paris to continue his education, which included mathematics, and then later to the University of Poitiers, where he received his degree in law in 1616. Law was the profession of his father and older brother, but he was not interested in it and became a soldier instead. In 1618 he went to Holland to serve as an officer in the army of the Prince of Orange. There he met Isaac Beeckman, a Dutch philosopher, mathematician and physicist, and a friendship developed which continued many years. Beeckman brought his friend up-to-date with mathematical developments,[2] then or later. The next year he served in the army of the Duke of Bavaria and was stationed at Ulm during the winter of 1619–1620. Here he met Johann Faulhaber, a German mathematician. He did not stay in Germany long and after some years of travel settled in Paris in 1626. After visiting Holland in 1628, he moved there in 1629 and remained until 1649 with only a few visits back to France. The *Discours de la Méthode*, including the *Géométrie*, was published in 1637. In September 1649 he went to Stockholm at the repeated invitation and urging of Queen Christina of Sweden. The climate and the conditions there proved too much for his health and he died February 11, 1650. The belongings of Descartes were taken care of by Chanut, the French ambassador at Stockholm and a friend of his. An inventory was made of manuscripts which were not personal, and this collection was released to Chanut by the heirs. Item M of the inventory reads, "M. Environ seize feuillets in octavo soubs ce titre [about sixteen leaves in octavo under this title]: Progymnasmata de solidorum elementis." In 1653 Chanut sent the manuscripts to Clerselier, his brother-in-law and also a friend of Descartes. The box containing the manuscripts was transhipped at Rouen to a boat going to Paris. When it reached Paris the boat was wrecked; the box of manuscripts fell into the river and was not recovered until three days later, at a distance from the site of the wreck. The papers had to be separated and hung to dry on cords in various rooms about the house.[3]

Clerselier published part of the papers, notably in three volumes of letters in 1657, 1659 and 1667 and in another work of 1664. Others remained unpublished, even though he had indicated in the preface of 1667 that a volume of fragments would be published. The unpublished papers were made available to others by Clerselier and references to a few of them appear in some contemporary works. Clerselier died in 1684.

Leibniz was in Paris in 1675–1676, and before leaving to enter the service of the House of Hanover (he was a lawyer by profession) he made copies of various unpublished manuscripts in the hands of Clerselier, including the *De Solidorum Elementis,* in 1676.

The original Descartes manuscript disappeared without being published, and has never been found. The Leibniz copy also disappeared and was unknown until it was found nearly 200 years after it had been made. It was published in 1860 after it was discovered by Comte Foucher de Careil among a collection of uncatalogued Leibniz papers in the Royal Library of Hanover. He was then actively seeking unpublished Descartes manuscripts, which he published in his *Oeuvres inédites de Descartes.*[4] An account of the manuscripts and his work in verifying them is given in the preface to that edition. While he translated other papers, he was unable to translate the *De Solidorum Elementis.*

The significance of the manuscript was immediately pointed out by E. Prouhet in a brief note in the *Comptes rendus*[5] (which I will refer to as Prouhet I). He followed this with a commentary and a French translation in the *Revue de l'Instruction publique*[6] which I will refer to as Prouhet II. C. Mallet, in a review of the Foucher de Careil collection,[7] had severely criticized the edition of the Latin text, which on its face was quite corrupt; it obviously contained omissions, incorrect readings of words, misplaced punctuation and passages with no meaning. Mallet questioned whether the corruptions were due to the manuscript itself or to bad printing and hasty proofreading; if the former, Foucher de Careil should at least have warned the reader and could even have supplied some notes. Several examples of unintelligible passages were given, with suggestions for their correction. However, Foucher de Careil was not a mathematician and deserves great credit for discovering and publishing the manuscript. Prouhet tried to straighten out what he could in his translation and commentary and did quite a good job with the inadequate material available to him. Some correspondence by Mallet and Prouhet, discussing several unintelligible and obscure passages, followed in the same journal.[8]

Thirty years later, in 1890, Vice-Admiral Ernest de Jonquières published a memoir containing a reprinting of the Latin text as published by Foucher de Careil, together with a "revised and completed" Latin text as he thought it ought to read, with a French translation, commentary and notes.[9] He knew of

Prouhet I but he was unaware of Prouhet II with its translation. While most of his readings agree with Prouhet, some do not, but he explained some things which Prouhet did not. His translation considerably amplifies the text.

In preparing to edit the *De Solidorum Elementis* for the standard edition of the collected works of Descartes, one of the editors, Charles Adam, traveled to Hanover in 1894 and studied the Leibniz manuscript together with his brother Henri. The preliminary text which they then established was later improved by further study of the manuscript by A. Meillereux, a pupil of Adam, in 1897, and particularly by J. Sire, who worked at Hanover, cataloguing Leibniz's papers for several years, thus becoming familiar with his handwriting, and revised Adam's text in 1906. The final result was published in 1908 in Descartes, *Oeuvres,* Vol. 10, pages 265–276, with a folding plate.[10] However, further inspection of the manuscript (which was sent to Nancy for the editors of the *Oeuvres*) revealed a number of errors in that edition: these were corrected in *Oeuvres,* Vol. 11, pages 690–692. In the second edition of Vol. 10 of the *Oeuvres,* pages 276 and 686–689, P. Costabel has added corrections based on his own examination of the manuscript.[11]

An Italian translation of the 1908 version was published in 1920 by A. Natucci. I have not seen any French translation of the corrected text, or any translation of it into any language other than Italian.

3 Description of the Leibniz Copy

The Leibniz copy of the Descartes manuscript is in the Niedersächsische Landesbibliothek at Hanover.[12] It is written on a double folio sheet folded in two, making four folio pages measuring a little over 20 × 30 cm (8 × 12 in.). Page 1, which starts with the title, is completely filled with writing, from very close to the top to 1 cm from the bottom. Page 2 begins an entirely different topic without any warning mark or word, with the page not entirely full; it has two small figures inserted in the text and a number of closely packed small tables. Page 3 continues the subject of page 2 and the writing occupies less than 60% of the sheet; it ends with a large closely packed table (which is printed on a folding plate in the *Oeuvres*). Page 4 is blank. The handwriting is small and the lines are crowded; in many places it is quite difficult to make out letters or words. A photographic copy of the manuscript is reproduced pp. 11–21.

In view of the fact that two different subjects are treated Prouhet and the *Oeuvres* divide the paper into two parts, Part I and Part II.

A note by P. Costabel in the second edition of the *Oeuvres*[13] states that it is not certain that Leibniz had entirely reproduced the original manuscript. There is evidence in the manuscript itself of possible omissions of a word or phrase here and there, a full sentence, and headings and repeated terms of tables, as will be seen in my commentary (Sections 6 and 11).

Nothing is known concerning the original manuscript except its title and that there were about sixteen octavo leaves. Since neither the size of these leaves is known, nor the size of the handwriting and spacing of the lines, nor the spacing of the tables, no comparison can be made with the known copy. But an inference that the copy contains practically all of the original is not precluded. If a sheet the same size as the folio pages of the Leibniz manuscript were folded twice to form octavo pages, their size would be approximately 10 × 15 cm (about 4 × 6 in.). An octavo of Descartes' time would probably be close to this size.[14] These are rather small sheets; with handwriting not as small, lines not as close, and tables not as crowded, as in the copy, the text of the Leibniz copy could reasonably have occupied most of the original sixteen octavo pages. I am inclined to the view that Leibniz copied nearly all of the original manuscript, with only a few incidental or accidental omissions. He had a number of manuscripts of Descartes and others and even, at a later date, considered the publication of a collection. It seems hardly likely that he would have taken liberties with the texts.

The transcription of the text of the manuscript follows as part of this section.

To sum up briefly what was stated above concerning the standard text in the *Oeuvres* of Descartes, the Latin text published in Vol. 10 in 1908 represented a great improvement over that of the *editio princeps,* Foucher de Careil's corrupt version of 1860. The editors made a number of corrections to the 1908 text in Vol. 11, published in 1909, and P. Costabel added some further corrections in his notes to the second edition of Vol. 10 in 1966. The result of combining the text as printed in *Oeuvres,* Vol. 10 with these two sets of corrections will be referred to as the *Oeuvres* text. The text given here, although it owes much to the latter, has been established on the basis of an independent reading of photographs of the manuscript, and adopts several readings which are different from the *Oeuvres* text. The reasons for the differences will become clear from the textual notes in this section and the commentary in Sections 6 and 11 (to which discussion of the meaning of the text is postponed).

As in the *Oeuvres,* the two parts of the manuscript are numbered I and II. In addition, the paragraphs have been numbered throughout for convenience of reference. There is little or no paragraphing in the manuscript, but the paragraphing here follows that of the *Oeuvres,* with a few changes. Notably, the order of Paragraphs 20 to 22 has been reversed from that of the manuscript and the *Oeuvres,* for the reason explained in Section 11, n. 111, p. 131. Also, Paragraphs 23 to 29 are somewhat confused in the manuscript and have been rearranged. In our arrangement each numbered paragraph except 28 consists of a description and a corresponding table; Paragraph 28 in the manuscript lacks the description, but this has been supplied by me, following Prouhet and de Jonquières. The rearrangement is shown by a table of the order in manuscript; 23T and 23D refer to the table and description in Paragraph 23 as numbered here, etc.

Table	Description
29T	blank
	23D written across the page
23T	written below on the left
24T	24D written side by side
25T	25D written side by side
26T	26D written side by side
27T	27D written side by side
28T	29D written side by side

As can be seen from this list, the first written table has been moved to the end and the last written description has been moved down to go with it, leaving the penultimate table (28) as the one without the description. The order here differs from that in the *Oeuvres* in that the last two paragraphs are interchanged. Thus our order matches that of the entries in the large table at the end of the manuscript.

Some annotations by Leibniz in the lower part of the left margin of Bl. 1 verso have been inserted (as Paragraphs 23a and 28a) next to that part of the text to which they refer. Paragraphs 33a and 34 are also annotations of his to the large table, the first written in the blank space in column 5, with a lead line going to the second formula in line 10, column 5 of the table,[15] the second written below the table.

In the large table (p. 28) the lines of the table have been numbered for convenience of reference.

In the text, following standard editorial conventions, we denote additions to the manuscript by angled brackets, and words to be deleted from the manuscript by square brackets. Emendations are indicated by italics. Our numbering of the paragraphs and lines of the tables is also done in italics. In the apparatus at the foot of the text 'cod.' refers to the Leibniz manuscript, 'Oeuvres' to the text of *Oeuvres de Descartes* (including the later corrections, as explained above).

On the following pages (11–21) are reproduced in facsimile those pages of the manuscript which have writing on them. First we give a reproduction of each of the three pages at about 50% of the size of the original and then again of each page (divided in thirds), the same size as the original.

Manuscript Page 1 (Folio 1 recto)

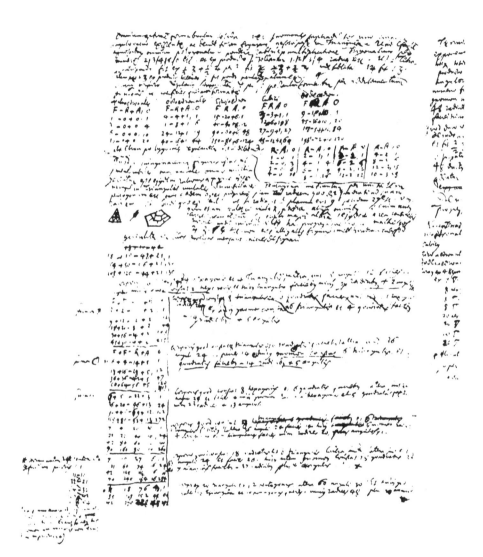

Manuscript Page 2 (Folio 1 verso)

Manuscript Page 3 (Folio 15 recto)

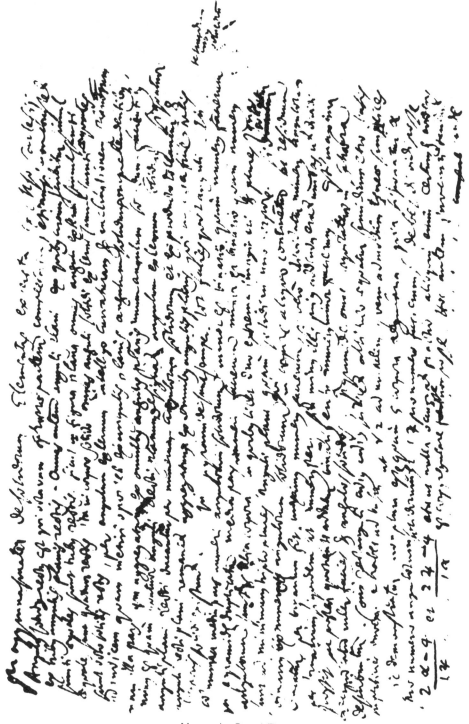

Manuscript Page 1 Top

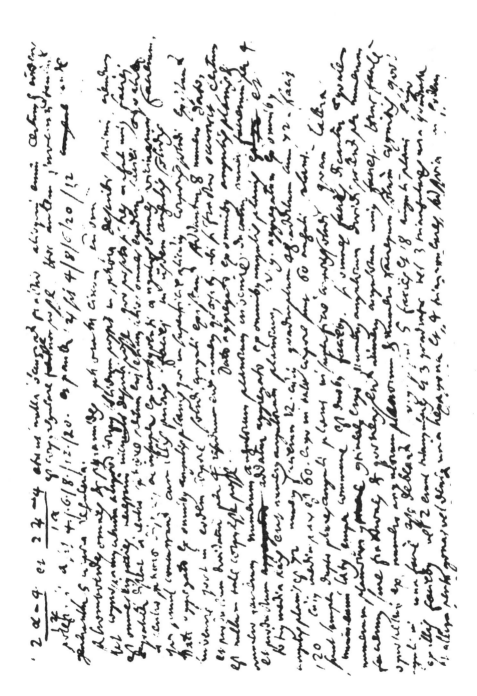

Manuscript Page 1 Middle

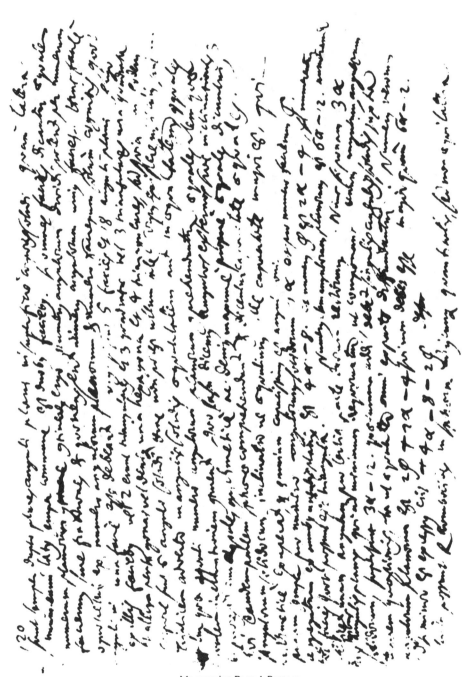

Manuscript Page 1 Bottom

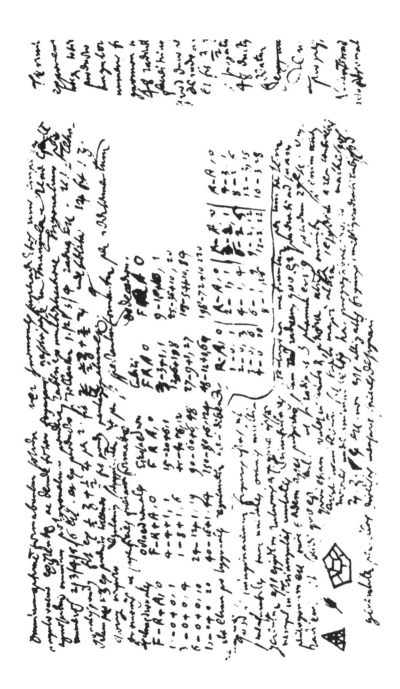

Manuscript Page 2 Top

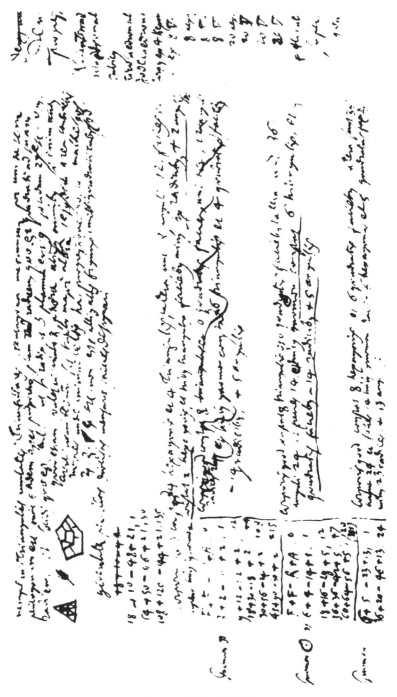

Manuscript Page 2 Middle

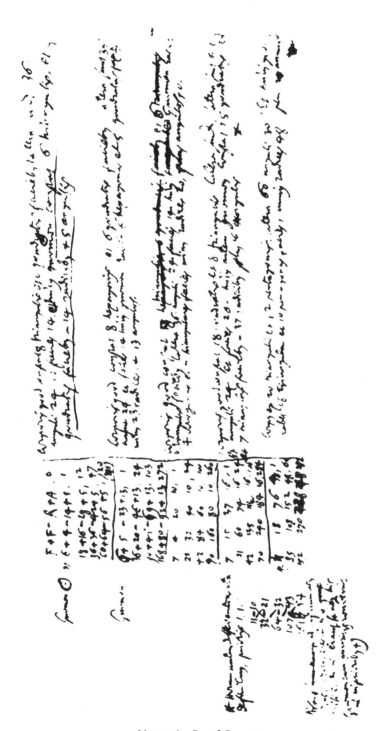

Manuscript Page 2 Bottom

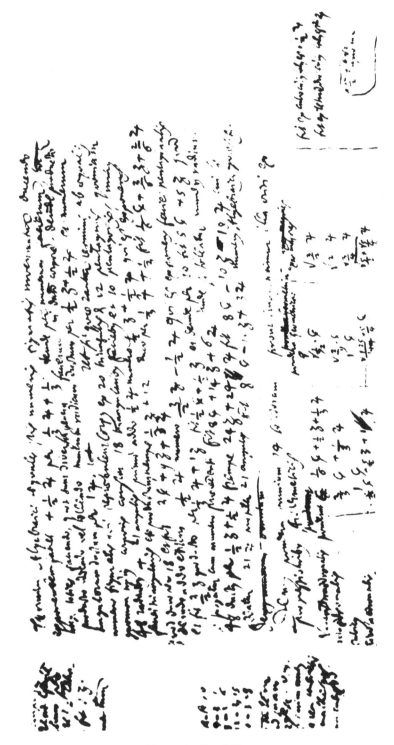

Manuscript Page 3 Top

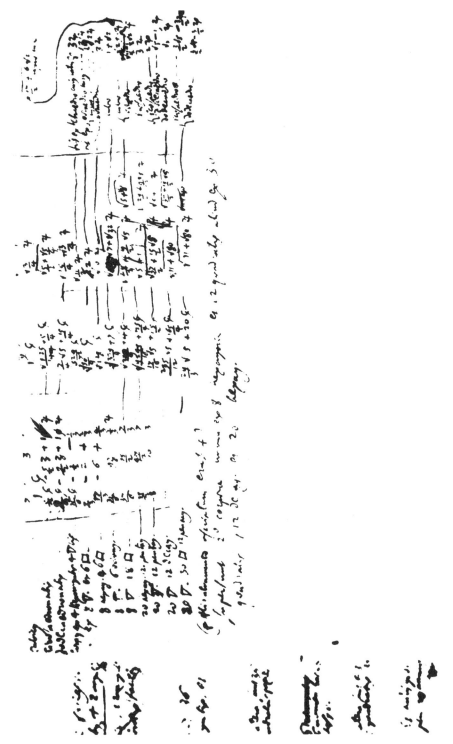

Manuscript Page 3 Middle

Progymnasmata de Solidorum Elementis excerpta ex Manuscripto Cartesii

1 Angulus solidus rectus est qui octavam sphaerae partem complectitur, etiamsi non constet ex tribus angulis planis rectis. Omnes autem anguli planorum, ex quibus circumscribitur, śimul sumti, aequales sunt tribus rectis.

2 Sicut in figura plana omnes anguli externi, simul sumti, aequales sunt quatuor rectis: ita in corpore solido omnes anguli solidi externi, simul sumti, aequales sunt octo solidis rectis. Per angulum externum intelligo curvaturam seu inclinationem planorum ad invicem, quam metiri oportet ex angulis planis angulum solidum comprehendentibus. Nam illa pars qua aggregatum ex omnibus angulis planis unum angulum solidum facientibus, minus est quam quatuor anguli recti planum ⟨facientes⟩, designat angulum externum solidum.

3 Si quatuor anguli plani recti ducantur per numerum angulorum solidorum et ex producto tollantur 8 anguli recti plani, remanet aggregatum ex omnibus angulis planis qui in superficie talis corporis solidi existunt.

4 In pyramide sunt semper tot facies quot anguli. In columna media pars numeri angulorum solidorum minor est binario quam numerus facierum. In pyramide duplicata media pars numeri facierum minor est binario quam numerus angulorum. Sunt et alia corpora in quibus licet duo extrema imaginari et plures zonas. Sunt ad minimum triplo plures anguli plani quam solidi in uno corpore. Si tollatur binarius ex numero angulorum solidorum qui in corpore aliquo continentur, et residuum ducatur per binarium, fit maximus numerus facierum. Si vero dividatur numerus angulorum per binarium, si quidem sit numerus par, sin minus illi prius addenda erat unitas

Title: Manuscripto: Mso. cod.
§1   planorum: plani cod., *corr.* Federico
§2   seu: & Oeuvres, *but* seu *is clear in cod.*   facientes *add.* Prouhet II
§4   erat: *sic* cod. erit Oeuvres   que *del.* Oeuvres

ut dividi possit, ac postea quotienti addatur binarius, erit[que] numerus minor facierum. Est maxima reciprocatio inter facies et angulos solidos.

4a tetraedron
cubus
octaedrum

5 Pyramides omnes aequilaterae in sphaera describuntur.

6 Coni rectanguli cuius [cuius] scilicet altitudo aequatur semidiametro basis, superficies convexa se habet ad basin ut $\sqrt{2}$ ad unitatem, quemadmodum lineae simplices.

7 Sic demonstratur non plura esse quam 5 corpora regularia: quia si ponatur α pro numero angulorum solidorum, et 1λ pro numero facierum, debet dividi posse

$$\frac{2\alpha - 4}{1\lambda} \quad \text{et} \quad \frac{2\lambda - 4}{1\alpha} \, ,$$

ita ut nulla occurrat fractio; alioquin enim certum et evidens est corpus regulare esse non posse. Hoc autem inveniri tantum potest, si α sit 4./.6./8./12./20, et pariter 1λ sit 4/8/6/20/12: unde generantur 5 corpora regularia.

8 Rhomboeides omnes et pyramides sphaeram circumscribunt.

9 Ut cognoscamus utrum aliquod corpus solidum possit in sphaera describi, primo sciendum est omnes eius facies necessario in circulo describi posse. Quo posito, si tres anguli unius faciei aequaliter distent a centro sphaerae, certum erit etiam alios omnes eiusdem faciei aequaliter a centro sphaerae distare; ac insuper ex consequenti, angulos omnes vicinarum facierum, qui simul concurrunt cum illis prioris faciei in iisdem angulis solidis.

10 Dato aggregato ex omnibus angulis planis qui in superficie alicuius corporis solidi existunt, invenire quot in eodem corpore solidi anguli existant. Addantur 8 numero dato, et productum dividatur per 4: residuum erit numerus quaesitus, ubi si fractio occurrat, certum est nullum tale corpus esse posse.

11 Dato aggregato ex omnibus angulis planis et numero facierum, numerum angulorum planorum invenire. Ducatur numerus facierum per 4, et productum addatur aggregato ex omnibus angulis planis: et totius media pars erit numerus angulorum planorum. V.g., aggregatum ex omnibus angulis planis est 72, numerus facierum 12, cuius quadruplum 48 additum cum 72 facit 120, cuius media pars est 60: ergo in tali corpore sunt 60 anguli plani.

12 Sunt semper duplo plures anguli plani in superficie corporis solidi quam latera; unum enim latus semper commune est duobus faciebus.

---

§6   cuius *del.* edd.

13 Si omnes facies dicantur aequalem numerum ⟨angulorum⟩ planorum continere, ergo numerus angulorum dividi poterit per numerum facierum sine fractione, et quotiens erit numerus angulorum unius faciei. Hinc facile cognoscetur, ex numero angulorum planorum et numero facierum solum cognitis, quot anguli in una facie esse debeant. V.g., si sint 5 facies et 18 anguli plani, ergo ex illis faciebus vel 2 erunt triangulares et 3 quadratae, vel 3 triangulares una quadrata et altera pentagona, vel denique una hexagona et 4 triangulares. Sed quia in eodem corpore sunt 6 anguli solidi, hinc non potest ullum tale corpus existere, nisi cuius sint...

14 Triplicem adverto in angulis solidis aequalitatem aut inaequalitatem: aequales dicuntur *qui* aequali numero angulorum planorum comprehenduntur; aequales item *qui* aequalem inclinationem continent, quo casu dicemus angulos externos sive inclinationis ⟨aequales esse⟩, et priores dicemus aequales arithmetice; ac denique maxime proprie aequales dicuntur, qui eandem partem sphaerae comprehendunt, et dicentur capacitate aequales.

15 Angulorum solidorum inclinatione aequalium ille capacitate maior est, qui arithmetice exuperat; et omnium capacissimus est angulus coni.

16 Ponam semper pro numero angulorum solidorum α et pro numero facierum ϕ. Aggregatum ex omnibus angulis planis est $4\alpha - 8$, et numerus ϕ est $2\alpha - 4$, si numerentur tot facies quot possunt esse triangula. Numerus item angulorum planorum est $6\alpha - 12$, numerando scilicet unum angulum pro tertia parte duorum rectorum. Nunc si ponam 3α pro tribus angulis planis qui ad minimum requiruntur ut componant unum angulum angulorum solidorum, supersunt $3\alpha - 12$, quae summa addi debet singulis angulis solidis iuxta tenorem quaestionis, ita ut aequaliter omni ex parte diffundantur. Numerus verorum angulorum planorum est $2\phi + 2\alpha - 4$, qui non debet esse maior quam $6\alpha - 12$; sed si minor est, excessus erit $+ 4\alpha - 8 - 2\phi$.

17 Describi possunt Rhomboeides in sphaera cuiuscumque quantitatis, sed non aequilatera.

18 Omnium optime formabuntur solida per gnomones superadditos uno semper angulo vacuo existente, ac deinde totam figuram resolvi posse in triangula. Unde facile agnoscitur omnium polygonalium pondera haberi ex multiplicatione trigonalium per numeros 2./3./4./5./6, etc., et ex producto si tollantur 1./2/3/4 radices, etc.

19 Ut: Tetragonalium pondus fit ex $\frac{1}{2}\mathcal{Z} + \frac{1}{2}\mathcal{X}$ per 2: fit $\frac{2}{2}\mathcal{Z} + \frac{2}{2}\mathcal{X}$, unde sublata $1\mathcal{X}$ fit $1\mathcal{Z}$; item per 3 ex producto tollendo $2\langle\mathcal{X}\rangle$, fit pondus pentagonalium, etc.

§13   angulorum *add.* Oeuvres. Cod. *has deleted word after* planorum
sint...*sic* cod., *indicating a lacuna;* sint 2 triangulares facies et 3 quadratae *suppl.* Prouhet II
§14   qui *(bis):* quae cod., *corr.* Oeuvres   aequales esse *suppl.* Prouhet II, de Jonquières
§19 $2\langle\mathcal{X}\rangle$ *suppl.* Federico

20 Ita etiam polygonales regulariter fieri debent:

R-A,	0	R-A,	0	R-A,	0	R-A,	0
1-0,	1	2-1,	1	3-2,	1	4-3,	1
2-0,	3	4-1,	4	6-2,	5	8-3,	6
3-0,	6	6-1,	9	9-2,	12	12-3,	15
4-0,	10	8-1,	16	12-2,	22	16-3,	28.

21 Quod si imaginaremur figuras istas ut mensurabiles, tunc unitates omnes intelligerentur esse eiusdem rationis ac figurae ipsae: nempe in triangulis unitates triangulares; pentagona metiuntur per unitatem pentagonam etc. Tunc eadem esset proportio plani ad radicem quae est quadrati ad suam radicem; et solidi quae est cubi: ut si radix sit 3, planum erit 9, solidum 27, etc., v.g. Quod etiam valet in circulo et sphaera aliisque omnibus. Si enim unius circuli circumferentia sit triplo maior altera, eiusdem area⟨m⟩ continebit novies. Unde animadvertis has progressiones nostrae matheseos, $\mathcal{2e}$, $\mathcal{z}$, $\mathcal{ce}$, etc., non esse alligatas figuris lineae, quadrati, cubi, sed generaliter per illas diversas mensurae species designari.

22 Quinque corpora regularia, simpliciter ut per se spectantur, formantur per additamentum gnomonis, ut superficies fuerant formatae:

Tetraedronales

F−R+A,	0
1− 0+0,	1
3− 0+0,	4
6− 0+0,	10
10− 0+0,	20

Cubici

F R A,	0
3− 3 +1,	1
12− 6 +1,	8
27− 9 +1,	27
48 −12 +1,	64

Octaedronales

F− R+A,	0
4− 4 +1,	1
12− 8 +1,	6
24 −12 +1,	19
40 −16 +1,	44

Eicosaedron

F R A,	0
15−20+6,	1
45−40+6,	12
90−60+6,	48
150−80+6,	124

Dodecedron

F R A,	0
9−18+10,	1
45−36+10,	20
108−54+10,	84
198−72+10,	220.

23 Corporis quod constat 4 hexagonis et 4 triangulis, latera sunt 18, anguli 12, facies 8. Igitur huius gnomon constat 2 hexagonis et tribus triangulis faciebus, minus sex radicibus, + 2 angulis:

§20   *Heading, 0:* Oeuvres *prints this, here and throughout* (§§ 22–24) *as a capital O. Sense demands that it be 'zero', and that is what is in the manuscript.*

§20,21,22   *In the manuscript these paragraphs appear in the order* §22, §21, §20. *Reversed by Federico. See Section* 11 n. 111, p. 131.

§21   area⟨m⟩: area cod., *corr.* Federico.

§22   Dodecedron. *sic* cod., Dodecadron Oeuvres

Gnomon

$$
\begin{array}{ll}
F+ \; F- \; R+A, & 0 \\
3+ \; 2- \; 6+2\,, & 1 \\
9+12-12+2\,, & 12 \\
18+30-18+2\,, & 44 \\
30+56-24+2\,, & 108 \\
45+90-30+2\,, & 215.
\end{array}
$$

23a Horum autem differentias ita definiemus, prioris 1, 1

$$
\begin{array}{ll}
11 & 10 \\
32 & 21 \\
64 & 32 \\
107 & 43 \\
161 & 54.
\end{array}
$$

24 Corporis quod constat 8 triangulis et 6 quadratis faciebus, latera sunt *24*, anguli *12* et facies 14. Et huius gnomon constat 6 triangulis et 4 quadratis faciebus, − 14 radicibus, + 5 angulis:

Gnomon

$$
\begin{array}{ll}
F+ \; F- \; R+A, & 0 \\
6+ \; 4-14+5, & 1 \\
18+16-28+5, & 12 \\
36+36-42+5, & 47 \\
60+64-56+5, & 120 \\
& (245).
\end{array}
$$

25 Corporis quod constat 8 hexagonis et 6 quadratis faciebus, latera sunt 36, anguli 24 et facies 14. Huius gnomon habet 6 hexagonas et 5 quadratas facies, minus 23 radices, + 13 angulos:

Gnomon

$$
\begin{array}{ll}
6+ \; 5-23+13, & 1 \\
36+20-46+13, & 24 \\
90+45-69+13, & 103 \\
168+80-92+13, & 272.
\end{array}
$$

26 Corporis quod constat 8 triangulis et 6 octangulis faciebus, latera 36, anguli 24, facies 14. Huius gnomon habet 4 octagonas et 7 triangulares facies, minus radices 20, plus angulos 10:

§24   24: 36 cod.; 12: 24 cod., *corr.* Prouhet II, de Jonquières

7	4	20	10,	1
21	32	40	10,	24
42	84	60	10,	100
70	160	80	10,	260.

27 Corporis quod constat 18 quadratis et 8 triangulis, latera sunt [latera sunt] 48 et anguli 24 et facies 26. Huius autem gnomon constat 15 quadratis et 7 triangulis faciebus, – 37 radicibus, plus 16 angulis:

7	15	37	16,	1
21	60	74	16,	24
42	135	111	16,	106
70	240	184	16,	284.

28 ⟨Corporis quod constat 12 pentagonis et 20 hexagonis faciebus, latera sunt 90, anguli 60 et facies 32. Huius gnomon habet 11 pentagonas et 18 hexagonas facies, minus 76 radices, plus 48 angulos:⟩

11	18	76	48,	1
55	108	152	48,	60
132	270	228	48,	282.

28a Qui ad sinistrum latus lineae characteres in Mso elisi et dubii erant. (Neque hic gnomon cum numeris convenit ut in prioribus.)

29 Corpus ex 20 triangulis et 12 pentagonis: latera 60, anguli 30, ⟨facies 32⟩, et huius gnomon habet 18 triangula⟨s⟩ et 10 pentagonas facies, minus radices 48, plus 21 angulis:

18 +	10 −	48 + 21,	1
54 +	50 −	96 + 21,	30
108 +	120 −	144 + 21,	135.

30 Termini algebraici aequales istis numeris figuratis inveniuntur ducendo exponentem faciei $+\frac{1}{2}\mathcal{X}$ per $\frac{1}{3}\mathcal{X}+\frac{1}{3}$, deinde per numerum facierum; hocque toties faciendo, quot sunt diversa genera facierum in dato corpore; deinde producto [producto] addendo vel tollendo numerum radicum ductum per$\frac{1}{2}\mathcal{Z}$ $+\frac{1}{2}\mathcal{X}$, et numerum angulorum ductum per $1\mathcal{X}$.

---

§28 *Suppl.* Federico, *following* Prouhet II; *similarly* de Jonquières
132: 152 cod., Oeuvres; *corr.* Federico
§29 facies 32 *suppl.* Prouhet II
triangula cod., triangulas Oeuvres
In the manuscript this table is after §21: transferred here by Prouhet II, de Jonquières

	[1]	[2]	[3] pondera geometrica	[4] axes maiores
1	Numerus tetraedronalis ponderat	$\frac{1}{6}\,\mathfrak{a} + \frac{1}{2}\,\mathfrak{z} + \frac{1}{3}\,\mathcal{H}$	$\sqrt{\tfrac{1}{72}}\,\mathfrak{a}$	$\sqrt{\tfrac{3}{2}}\,\mathcal{H}$
2	octaedronalis	$\frac{2}{3}\,\mathfrak{a} + \frac{1}{3}\,\mathcal{H}$	$\sqrt{\tfrac{2}{9}}\,\mathfrak{a}$	$\sqrt{2}\,\mathcal{H}$
3	cubicus	$1\,\mathfrak{a}$	$1\,\mathfrak{a}$	$\sqrt{3}\,\mathcal{H}$
4	eicosaedronalis	$\frac{5}{2}\,\mathfrak{a} - \frac{5}{2}\,\mathfrak{z} + 1\,\mathcal{H}$	$\sqrt{\tfrac{125}{144}} + \frac{5}{4}\,\mathfrak{a}$	$\sqrt{\tfrac{5}{2} + \sqrt{\tfrac{5}{4}}}\,\mathcal{H}$
5	dodecaedronalis	$\frac{9}{2}\,\mathfrak{a} - \frac{9}{2}\,\mathfrak{z} + 1\,\mathcal{H}$	$\frac{7}{4}\sqrt{5} + \frac{15}{4}\,\mathfrak{a}$	$\sqrt{\tfrac{15}{4} + \sqrt{\tfrac{3}{4}}}\,\mathcal{H}$
6	Corpus ex 4 hexangulis et 4 ▽lis	$\frac{11}{6}\,\mathfrak{a} - \frac{1}{2} - \frac{1}{3}$	$\sqrt{\tfrac{529}{72}}\,\mathfrak{a}$	$\sqrt{\tfrac{11}{2}}\,\mathcal{H}$
7	ex 8▽ et 6□	$\frac{7}{2} - 2 + \frac{2}{3}$	$\sqrt{\tfrac{50}{9}}\,\mathfrak{a}$	$2\,\mathcal{H}$
8	8 hexag. et 6□	$\frac{17}{3} - 6 + \frac{4}{3}\,\mathcal{H}$	$\sqrt{128}\,\mathfrak{a}$	$\sqrt{10}\,\mathcal{H}$
9	8▽, 6 octag.	$\frac{31}{6}\quad \frac{9}{2}\quad -\frac{1}{3}\,\mathcal{H}$	$\sqrt{\tfrac{392}{9}} + 7\,\mathfrak{a}$	$\sqrt{7 + \sqrt{32}}\,\mathcal{H}$
10	8▽ 18□	$\frac{37}{6}\quad \frac{15}{2}\quad \frac{7}{3}$	$\sqrt{\tfrac{200}{9}} + 4\,\mathfrak{a}$	$\sqrt{5 + \sqrt{8}}\,\mathcal{H}$
11	20 hexag. 12 pentag.	$\frac{35}{2}\quad \frac{47}{2}\quad 7$	$\sqrt{\tfrac{9245}{16}} + \frac{125}{4}\,\mathfrak{a}$	$\sqrt{\tfrac{29}{2} + \frac{9}{2}\sqrt{5}}\,\mathcal{H}$
12	20▽, 12 pentag.	$8\quad 10\quad 3$	$\frac{17}{6}\sqrt{5} + \frac{15}{2}\,\mathfrak{a}$	$\sqrt{5 + 1}\,\mathcal{H}$
13	20▽, 12 decag.		$\frac{235}{12}\sqrt{5} + \frac{165}{4}\,\mathfrak{a}$	$\sqrt{\tfrac{37}{2} + \frac{15}{2}\sqrt{5}}\,\mathcal{H}$
14	20▽, 30□, 12 pentag.		$\frac{29}{3}\sqrt{5} + 20\,\mathfrak{a}$	$\sqrt{11 + \sqrt{80}}\,\mathcal{H}$

[5]

fit ex cubo cuius latus est $\sqrt{\tfrac{1}{2}}\,\mathcal{H}$　—　$2\,\mathcal{H}$

fit ex tetraedro cuius latus est $2\,\mathcal{H}$　—　$\sqrt{2}\,\mathcal{H}$; $\sqrt{3}\,\mathcal{H}$

fit ex tetraedro cuius latus est $3\,\mathcal{H}$

fit ex
- octaedro cuius — cubo : $1 + \sqrt{2}$
- cubo — octaedro : $\sqrt{2} + 1\,\mathcal{H}$

- { octaedro — icosahedro : $\sqrt{\tfrac{11}{2} + \frac{6}{2}\sqrt{2}}\,\mathcal{H}$; $3\,\mathcal{H}$
  icosahedro : $2\,\mathcal{H}$
- { dodecahedro : $\sqrt{5} - 1\,\mathcal{H}$
  dodecaedro : $\sqrt{5}\,\mathcal{H}$
- { icosaedro : $\frac{3}{2}\sqrt{5} - \frac{1}{2}\,\mathcal{H}$
  dodecaedro : $\frac{3}{2}\sqrt{5} - \frac{3}{2}\,\mathcal{H}$

Line 3 col. 4 $\sqrt{3}$: $\sqrt{\tfrac{3}{4}}$ cod., *corr.* de Jonquières

Line 6 col. 4 $\frac{11}{2}$ *correctly* cod., $\frac{11}{3}$ Foucher de Careil, Oeuvres

Line 9 col. 3 $3\frac{329}{9}$ cod., *corr.* de Jonquières

Line 10 col. 5 $\sqrt{\tfrac{11}{2} + \frac{1}{2}\sqrt{2}}$ cod., *corr.* Toomer (see p. 123 n. 15)

Line 12 col. 3 $3\frac{17}{16}$ cod., *corr.* Federico, following de Jonquières

31 Ut si quaerantur termini *adaequales* numeris figuratis qui repraesentent corpus ex 20 triangulis et 12 pentagonis, quoniam gnomon huius corporis constat 18 triangularibus faciebus et 10 pentagonis, minus 48 radicibus, +21 angulis, primo addo $\frac{1}{2}\mathcal{X}$ numero $\frac{1}{2}\mathcal{Z}+\frac{1}{2}\mathcal{X}$, qui est exponens faciei triangularis, et productum, nempe $\frac{1}{2}\mathcal{Z}+1\mathcal{X}$, duco per $\frac{1}{3}\mathcal{X}+\frac{1}{3}$: fit $\frac{1}{6}\mathcal{C}+\frac{3}{6}\mathcal{Z}+\frac{2}{6}\mathcal{X}$, quod duco per 18 et fit $3\mathcal{C}+9\mathcal{Z}+6\mathcal{X}$.

32 Deinde addo etiam $\frac{1}{2}\mathcal{X}$ numero $\frac{3}{2}\mathcal{Z}-\frac{1}{2}\mathcal{X}$, qui est exponens faciei pentagonalis, et fit $\frac{3}{2}\mathcal{Z}$, quo ducto per $\frac{1}{3}\mathcal{X}+\frac{1}{3}$, fit $\frac{1}{2}\mathcal{C}+\frac{1}{2}\mathcal{Z}$; et deinde per 10, fit $5\mathcal{C}+5\mathcal{Z}$; quod si iungatur cum numero praecedenti, fit $8\mathcal{C}+14\mathcal{Z}+6\mathcal{X}$. Unde si tollatur numerus radicum 48 ductus per $\frac{1}{2}\mathcal{Z}+\frac{1}{2}\mathcal{X}$, nempe $24\mathcal{Z}+24\mathcal{X}$, fit $8\mathcal{C}-10\mathcal{Z}-18\mathcal{X}$; cui si addatur $21\mathcal{X}$ propter 21 angulos, fit $8\mathcal{C}-10\mathcal{Z}+3\mathcal{X}$, numerus algebraicus quaesitus.

33 Denique pondera omnium 14 solidorum prout imaginamur illa oriri ex progressionibus arithmeticis:

(Large table: see p. 28)

33a $\sqrt{\frac{17}{2}-6\sqrt{2}}$
nescio cur

34 (Alio atramento ascriptum erat) Supersunt duo corpora, unum ex ⟨6 octogonis⟩, 8 hexagonis et 12 quadratis, aliud ex 30 quadratis, 12 decag. et 20 hexag.

§31   adaequales: abaequales cod., corr. Oeuvres
6$\mathcal{X}$: *sic* cod., *correctly;* 'on lit 5 ou 6', Oeuvres X p. 692
§32 $\frac{1}{3}\mathcal{X}+\frac{1}{3}:\frac{1}{3}\mathcal{X}+1\mathcal{Z}$ cod., *corr.* de Jonquières
§34   6 octogonis *suppl.* Prouhet II

4 Date of the Original Descartes Manuscript

While the date of 1676 for the Leibniz copy is known, the date of the manuscript of Descartes is not known.

Milhaud ascribes the manuscript to the winter of 1619-1620 as a "very probable" date on the basis of the contents of Part II.[16] This part is concerned with determining the formulas for polyhedral numbers, analogues of the figurate numbers in the plane. The large table in Part II of the manuscript gives the formulas, and other data, for the polyhedral numbers corresponding to fourteen polyhedra, the five regular solids and nine semiregular solids. Descartes called the formula the weight (pondus) and the number along one side of a face the "radix." The latter was the base for the powers in the formulas.

The German mathematician Faulhaber, whom Descartes met at Ulm, was a devotee of figurate numbers. He had published a work on the subject with the title "Numerus figuratus sive Arithmetica analytica arte mirabili inaudita nova constans," in 1614. According to Milhaud, Faulhaber had a table with 6 polyhedral numbers and uses the same terms, "pondus" and "radix," used by Descartes. It is clear that Descartes had a high opinion of Faulhaber.

As has been stated on page 5, Descartes was at Ulm during the winter of 1619-1620. This was an important period in his life; he did a great deal of thinking then, the nature of which is recounted in, and forms the basis of, the *Méthode*. He met Faulhaber and undoubtedly his own work on polyhedral numbers was prompted or suggested to him by Faulhaber or Faulhaber's work. Milhaud further states that the resemblance in the terminology used by both men is sufficient to fix an approximate date. He asks, "Is it not reasonable that, from their first meetings, Faulhaber, full of enthusiasm about the subject of his works, would have talked to Descartes about it, and that the latter would have concerned himself with the same problems?" But not much weight can be given to both using the word "radix," as this term was commonly used for the unknown quantity in an algebraic expression. Furthermore, Milhaud had a misconception of the content of Faulhaber's work.[17]

But this reasoning only gives an earliest possible date; the manuscript may well have been written at a later time. The *Oeuvres* include the manuscript in a group of "Opuscules de 1619-21." No reason is given but presumably the editors followed the same line of argument as Milhaud did later.

A date before which the manuscript was (in all probability) written can be derived from the manuscript itself. Three peculiar symbols appear which may be seen in our reproduction of the manuscript, p. 17. The printed forms of these symbols as given by Clavius[18] are 𝒙, 𝟛, 𝒄𝒄. Foucher de Careil did not know what they were and printed the first one as 4 (and sometimes as a 2), the second as 3 and the third one as 4; his transcriptions are close to the cursive forms in the manuscript. Prouhet II correctly surmised that they were "cossic" characters representing the unknown or the quantity being considered, its square, and its cube respectively. De Jonquières, who had not seen Prouhet II and worked only with the Foucher de Careil text, stated that the numbers were secret signs utilized by Descartes to conceal the information, comparable to the practice of a century earlier of some mathematicians announcing a discovery by a cryptic statement or cipher; he himself had fathomed their significance. Finally Charles Adam and his collaborators, in their investigations devoted to establishing the text for the *Oeuvres,* confirmed that these were indeed cossic symbols like those of the 1608 *Algebra* of Clavius; they were also found in other early manuscripts of Descartes.[19]

The first half of the 17th century was a period of transition from literal algebra to symbolic algebra. In the former the unknown quantity and its powers, and the operations in an equation would be stated in words, or abbreviations, or initial letters of the names. Thus the unknown would be referred to as "cosa" (the thing) or "coss" (in German and English, hence algebra was the "cossic art"), its square as "zenso" or "zensus," and its cube as "cubus" (there were variations, and the unknown was also referred to as "radix" or "res"). The symbols which have been mentioned appear to have evolved from the initial letters of the words. The old system in its advanced stage is in the textbook on algebra of 1608 by Clavius. Clavius was a Jesuit teacher and his textbooks were popular, particularly in Latin schools; Descartes no doubt first learned his algebra from it at the Jesuit school at La Flèche.

Changes were introduced by François Viète, notably the use of letters for quantities, in his *In artem analyticam isagoge* of 1591, but this was privately printed with limited distribution and not widely known until later; Viète's works were reprinted in 1646. Descartes had abandoned the old system by 1637 and his *Géométrie* introduced further improvements in notation. Where Viète used capital letters, with the vowels for unknowns and consonants for known (unspecified) quantities, Descartes used small letters, with the first part of the alphabet for known quantities and x, y, z for the unknowns. Notable was his introduction of numerical (arabic) exponents to indicate powers. Where Viète might write AAA for the cube of A, Descartes would write simply a^3, but he would write aa as well as a^2.[20]

The above indicates that the manuscript was most probably written before 1637. Descartes is said to have learned the then "modern" system from his friend Beeckman in Holland.[21] If so, this could have been either during his brief sojourn in Holland in 1618 or after he had settled there in 1629. If the former, which is quite unlikely, the continued use of cossic symbols after 1618 would need explanation.

When the text of the Leibniz copy was being established for the *Oeuvres*, G. Eneström was consulted in connection with the symbols.[22] He established their significance from the *Algebra* of Clavius, from which, he stated, Descartes had learned their meaning, rather than from Faulhaber,[23] and called attention to two Descartes items which used cossic symbols, taken from Isaac Beeckman's Journal: a letter dated March 16, 1619[24] and a note of October 1628.[25] The latter is a memorandum written down by Beeckman of some algebra communicated to him by Descartes. Cossic symbols are used throughout, and since they were used by Beeckman in reporting or copying Descartes in 1628 they evidently were also being used by Descartes at that time. This indicates that the manuscript could have been written in 1628 or later.

The subject matter of Part I suggests a *terminus a quo* later than the date proposed by Milhaud. The first and basic proposition concerns polar (exterior) solid angles, and polar spherical triangles and polygons are also utilized. The manner of introducing these and the incompleteness or lack of description would indicate that Descartes was referring to things already known (see p. 65). But polar solid angles were not clearly described before the 1627 work of Snell.[26] Also, various propositions relate to or necessitate knowledge of the area of spherical triangles and polygons and the measure of solid angles. These were introduced by Girard in 1629.[27] Both of these authors were living in Holland and their works were published in Holland. If Descartes derived his knowledge from the work of Snell and Girard, then Part I of the manuscript could not have been written before their work, and the date would probably be 1629 or later. I suggest circa 1630 as a possible date, considering that Descartes may still have been using cossic symbols up to that time.[28]

Part Two
Solid Geometry: The Elements of Solids

5 Some Geometric Background

This section supplies the necessary basis for statements made and things referred to in the following sections. It introduces and uses some of the terms used by Descartes so that their significance may become clear. The use of terminology and ideas not in the manuscript and not current at the time of the manuscript is avoided as far as possible. Except where otherwise indicated or obvious from the context, the discussion is limited to things which, it is believed, would have been known generally to mathematicians at the time of the manuscript, and to explanations of these things. (In the explanations and derivations we do not limit ourselves to seventeenth century terminology).

(1) *Polygons.* Descartes does not use the term "polygon" but uses "plane figure." Euclid (I def. 19)[29] used "rectilinear figures," defined as those contained by straight lines, and distinguished as trilateral, quadrilateral and multilateral. It is assumed that convex polygons are meant. The elements of the plane figure are the sides and the angles, which are equal in number. "Angle" is taken to be the complex of indefinite portions from the meeting point of the two lines and the portion of the plane between them—the sector of the plane in the neighborhood of the point.

(2) *Plane angles.* The size of a plane angle is determined by the amount of bending (turning) or inclination of the two lines with respect to each other. It is measured according to the arc intercepted on a unit circle. In Fig. 1, line OB, starting from an initial position of coincidence with line OA, is rotated about the point O to the position OB'; angle AOB' is measured by the arc AB' in relation to the full circumference of the circle. If the line OB is further turned to position OB'', perpendicular to OA, the angle AOB'' is a right angle, the fourth part of the circle; this is the unit of angle measurement used in geometry. Continuing the turning to position OB''', with AOB''' a straight line, we have a straight angle, measuring two right angles. Rotation through the full circle gives the four right angles of the plane.

The modern measure is the length of the intercepted arc on a unit circle. The angle measures one radian when this length is one unit (equal to the

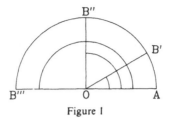
Figure 1

radius). The right angle is $\pi/2$ radians, the straight angle π radians and the four right angles of the plane 2π radians, the circumference of the unit circle. In trigonometry, however, degrees have been used since ancient times. The circumference is divided into 360 equal parts and the angle formed by lines from the center to the ends of one of these parts is one degree.

In what follows the letter $\triangle$ will be used in formulas to designate right angles. In the different measures: $1\triangle = \pi/2$ radians $= 90°$.

(3) *Polygon interior angle sum.* Proposition I, 32 of Euclid states that the sum of the three interior angles of a triangle is equal to two right angles. Euclid did not extend this; the extension to any convex polygon was given by Proclus (A.D. 410-485) in his commentary on Euclid. If the polygon has n sides or angles, the sum of the interior angles is $2(n-2)\triangle$. The proof given by Proclus[30] is evident from Fig. 2. From the point A of one angle lines are drawn to each of the points of the other angles with which it is not already connected. This divides the figure into triangles and the number of triangles in every case is two less than the number of sides of the figure. The sum of all the angles of all the triangles is equal to the sum of all the interior angles of the figure. Since the sum of the angles of each triangle is equal to two right angles, the sum of the interior angles is equal to the number of sides less two, times two right angles. This sum can be written as $2(n-2)\triangle$, where n is the number of sides or angles.

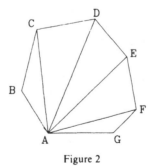
Figure 2

(4) *Exterior plane angles.* In Propositions 16 and 32 of Book I, Euclid produces (extends) one side of a triangle to form an "exterior" angle (literally an "outside" angle). The sum of the two angles on this extended line is two right angles: see Fig. 3(a), in which $\alpha + \beta = 2\triangle$. Two angles whose measures

add up to two right angles are supplementary, either angle being the supplement of the other, but this terminology was not used by Euclid.

Figure 3 shows three positions of the supplement of a given angle α. In (b), CO is drawn perpendicular to OB at O and DO perpendicular to OA, forming angle COD or β. The sum of the four angles about O is the four right angles of the plane, $4\triangle$; since two of them are right angles the remaining two, α and β, add up to two right angles, $2\triangle$. In (c) a point P is taken inside the angle AOB and perpendiculars PC and PD drawn to OB and OA, respectively, forming angle CPD. The four interior angles of the quadrilateral thus formed add up to $4\triangle$; two of these are right angles and hence the remaining two, α and β, add up to two right angles. These figures are used later for analogy. [The antiquity of forms (b) and (c) has not been ascertained.]

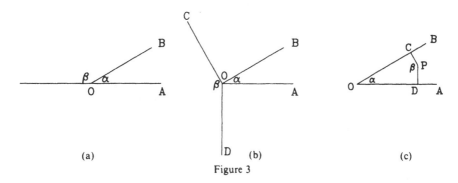

(a)   D   (b)   (c)

Figure 3

(5) *Exterior polygon angle sum.* Proclus also gives the sum of the exterior angles of a polygon.[31] His proof is quite simple and is illustrated by Fig. 4(a). The sum of one angle and its corresponding exterior angle is equal to two right angles, and the sum of all the exterior angles and all the interior angles together is equal to the number of sides (angles) of the figure times two right angles $(2n\triangle)$. The sum of the interior angles is equal to the number of sides less two, times two right angles $(2(n-2)\triangle)$. Hence the sum of the exterior angles is the difference, which is four right angles, $4\triangle$. Thus, no matter what the number of sides of a convex polygon, the sum of the exterior angles is the same, $4\triangle(2\pi,360°)$, the four right angles of the plane.

Figure 4(b) shows a different method of illustrating the exterior angles of the polygon, corresponding to that shown in Fig. 3(b). Proof of the sum is the same as stated above. This form is in Pólya.[32] He indicates that the exterior angles can be shifted so as to have the same vertex and fill the plane about this point, thus totalling 2π. This can be done because facing sides of adjacent exterior angles are parallel to each other and hence one angle can be moved without turning so as to have one side in common with another angle.

In Fig. 4(c) a point P is chosen inside the polygon and perpendiculars drawn to each of the sides. As in Fig. 3(c), each of the n angles about P so formed corresponds to the exterior angle of one of the respective n angles of the polygon; they cover the whole plane about the point without overlapping and hence their sum is 4△. (This form has not yet been found and is introduced here for analogy with a statement in the next section, p. 44.)

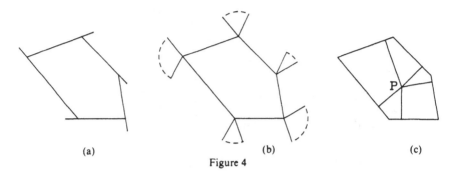

(a)   (b)   (c)

Figure 4

(6) *Polyhedra.* Descartes does not use the word polyhedron but refers to polyhedra as solid bodies, solids, or bodies. Undoubtedly he had in mind only convex bodies. Euclid, in defining equal and similar solid figures (XI def. 9, 10), refers to "solid figures" as being contained by planes. Heath discusses these definitions, as well as a variation of Heron,[33] and concludes that they are correct only if understood as applying to convex polyhedra.

The elements of the solid bodies of Descartes are the solid angles and the faces, the former including its plane angles and angles of inclination (see next paragraph) and the latter its plane angles and sides. The faces are "plane figures," namely, polygons, which are convex. "Solid angle" is taken as referring to the complex formed by indefinite portions of the lines and planes meeting at a point and the portion of space between them—the portion of the solid in the neighborhood of the point.

(7) *Inclination of planes.* The configuration formed by two half-planes meeting in a straight line is now called a dihedral angle. Euclid (XI def. 6) refers to the inclination of one plane with respect to another; this inclination is to be measured by the plane angle formed by a line in each plane perpendicular to the common line (the line in which the two planes intersect), at the same point, that is the angle formed by the intersections with the two planes of a plane perpendicular to the common line. In discussing solid angles, Descartes refers to the inclinations of adjacent planes and to the angle of inclination of these planes.

(8) *Solid angles.* A solid angle is formed by three or more planes meeting at a point. Euclid's second definition (XI def. 11) is "A solid angle is that which is contained by more than two plane angles which are not in the same plane and are constructed to one point." Figure 5 shows a solid angle OABC with three plane angles. There are three face angles, AOB, AOC and BOC, which Descartes refers to as the plane or face angles of the solid angle; these are measured by the great circle arcs intercepted on the unit sphere, AB, AC, and BC respectively in the figure, the sides c, b, a of the spherical triangle ABC.

There are three angles of inclination of the faces, which are measured as indicated in the preceding paragraph; these are also the angles α, β and γ of the spherical triangle ABC.

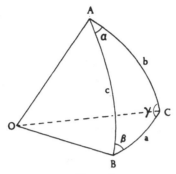

Figure 5

A solid angle is measured according to the area of the spherical polygon intercepted on the surface of the unit sphere; in the figure, the area of the spherical triangle ABC measures the solid angle OABC. With unit radius, the area of a spherical triangle is equal to the sum of its three angles (the inclination angles) less two right angles, in the figure, $\alpha + \beta + \gamma - 2\Delta$. This formula, the spherical excess formula, was given by Girard.[34] A spherical polygon with n sides can be divided into $n-2$ spherical triangles, and its area, the measure of the corresponding solid angle, obtained by adding together the areas of these triangles, in the same manner as Proclus did for the plane.[35] The result is that the area of the spherical polygon is equal to the sum of its angles less $2(n-2)$ right angles, $\alpha + \beta + \gamma + \delta + \varepsilon \cdots - 2\,(n-2)\Delta$.

If the three planes of Fig. 5 are mutually perpendicular, the result is a solid angle, an octant of the sphere. Each face angle is a right angle, as is also each inclination angle of the planes. The area intercepted on the sphere is the sum of the latter three right angles less two right angles, which comes to one right angle, $1\Delta\ (\pi/2)$. This is the eighth part of the area of a unit sphere (which is 8Δ, or 4π), and is the unit for solid angles. As it is the area which is the measure, a solid angle need not be trirectangular to measure one unit, one solid right angle, but may be any shape and have any number of faces as long as the intercepted area is 1Δ.

If the plane (face) angles of a solid angle are increased in size, the solid angle becomes more blunt and the intercepted area on the sphere becomes larger; the sum of the plane angles has thereby increased. With continuing increase of the plane angles, the solid angle becomes flatter, the sum of the plane angles is closer to the four right angles of the plane (4Δ), and the measure of the solid angle is closer to the area of a hemisphere (4Δ). When the solid angle has become completely flat, the sum of the face angles has become 4Δ, which is also the measure of the solid angle. Descartes refers to the difference between the four right angles of the plane and the sum of the face angles as the inclination of the solid angle. It measures, in a sense, the deviation of the solid angle from the plane. Euclid's proposition XI 21 states that the sum of the plane angles of a solid angle is less than four right angles (which is true only of convex solid angles); the difference is the inclination of the solid angle. This is also sometimes referred to as the "deficiency" of the solid angle.

(9) *Exterior solid angle.* The exterior solid angle of a given angle is the polar or supplemental angle. One manner of construction (corresponding to Fig. 3(b)) can be illustrated by means of Fig. 5. At the point O, construct a line perpendicular to the plane of OAB, and also lines at O perpendicular to the planes of BOC and COA, respectively. These three lines form a solid angle which is the supplement or polar of OABC. In other words, construct planes passing through the point O and perpendicular to each of the three lines OA, OB and OC, respectively. These planes meet in the three lines of the first construction and form the same solid angle.

The properties of the pair of solid angles is more easily seen by considering the polygons intercepted on the unit sphere. In Fig. 6, ABC is the spherical

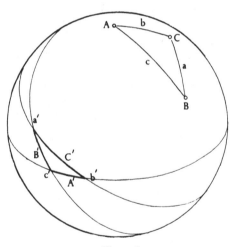

Figure 6

triangle intercepted on the sphere with center O (not shown, in the interior of the sphere) by the solid angle OABC. Letters A, B, C are also used for the angles of the triangle, the inclination angles of the planes. The sides of the triangle are a, b, c, equal to the face angles of the solid angle. With point A as pole construct its polar great circle (equator); that is, a plane is drawn through O, the center of the sphere, perpendicular to the line OA, and the intersection of this plane with the sphere is a great circle with pole A. Arc b'c' is part of this great circle. Similarly, the polar great circle of point B gives arc a'c' and that of point C gives arc a'b'. Triangle a'b'c' formed by these three arcs is the polar triangle of triangle ABC and its solid angle Oa'b'c' the polar (supplement, exterior) solid angle of OABC. (If the entire great circles are considered, they intersect in a second representation of the polar, but in reverse order, on the other side of the sphere, corresponding to the position of the polar solid angle in the constructions previously described.) It is easy to show by ordinary methods of Euclidian proof that sides A', B', C' of the polar are supplements of angles A, B, C, respectively, and that angles a', b', c' of the polar are supplements of sides a, b, c, respectively. Otherwise expressed, the relationship is that the plane (face) angles of each are supplements respectively of the inclination angles of the other, and the inclination angles of each are supplements respectively of the face angles of the other. The reciprocal relations can be written, referring to Fig. 6:

$$A + A' = 2\Delta, \ a + a' = 2\Delta,$$
$$B + B' = 2\Delta, \ b + b' = 2\Delta,$$
$$C + C' = 2\Delta, \ c + c' = 2\Delta,$$

The measure M of solid angle OABC, the area of triangle ABC, is $A+B+C-2\Delta$ and the measure M' of solid angle Oa'b'c' is $a'+b'+c'-2\Delta$. By using the preceding relations, one may obtain:

$$M = 4\Delta - (A'+B'+C')$$

and

$$M' = 4\Delta - (a+b+c).$$

Thus, the measure of the polar, exterior, solid angle of a given angle is equal to 4Δ, the four right angles of the plane, less the sum of the face angles of the given angle (and reciprocally).

Another way of showing the polar of a solid angle (corresponding to Fig. 3(c)) is illustrated by Fig. 7. In describing the formation of a polar, with respect to Fig. 5, we drew perpendiculars from the point O to each face. In Fig. 7 the perpendiculars to each face are drawn from a point P in the interior of

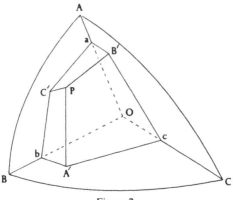

Figure 7

solid angle OABC. PA' is drawn perpendicular to the plane of face BOC and PB' perpendicular to AOC. The plane of A'PB' is perpendicular to the line OC and its intersections with the planes AOC and BOC are the lines B'c and A'c, respectively; therefore, angle A'cB' is the inclination angle of these two planes. The quadrilateral PA'cB' has two right angles, at A' and B', and the inclination angle at c; thus angle A'PB' is the supplement of the inclination angle. Now consider the perpendicular PC' to the face AOB; the angles C'PB' and C'PA' are the supplements of the other two inclination angles, as angle A'PB' is the supplement of the inclination angle mentioned above. Thus the solid angle formed by perpendiculars from P to each face of the given angle OABC is its polar angle.

The trihedral solid angle and spherical triangle have been used as examples for simplicity; the results are the same for any solid angle and the corresponding spherical polygon, provided of course that only convex solid angles are considered.

6 Translation and Commentary, Part I

Progymnasmata de Solidorum Elementis
excerpta ex Manuscripto Cartesii

1 Angulus solidus rectus est qui octavam sphaerae partem complectitur, etiamsi non constet ex tribus angulis planis rectis. Omnes autem anguli planorum, ex quibus circumscribitur, simul sumti, aequales sunt tribus rectis.

Preliminary Exercises on the Elements of Solids
Taken from a Manuscript of Descartes

1 A solid right angle is one which embraces the eighth part of the sphere, even though it is not formed by three plane right angles. But all the angles of the planes by which it is bounded, taken together, equal three right angles.

Comments. The first sentence defines the unit of measurement of solid angles, the solid right angle, which intercepts on a sphere one-eighth of its total area. The octant of the sphere, formed by three mutually perpendicular planes with each face angle being a plane right angle, is the unit solid angle. But a solid angle measuring one unit need not be formed from three plane right angles; it may be formed from three plane angles, not necessarily right, or from more than three plane angles, as long as the area intercepted on the sphere is one-eighth of the total area. See Section 5, paragraph 8 (p. 39).

In the second sentence the manuscript has "anguli plani," meaning "all the plane angles." It is of course false that all the face angles bounding a solid right angle add up to three right angles.[37] Therefore "plani" has been emended to "planorum," and the phrase translated as "the angles of the planes," which is taken to mean the angles the planes make with each other, the inclination angles (dihedral angles). Since the measure of the solid angle is the sum of these angles less two right angles, their sum must be equal to three right angles if the solid angle is to measure one right angle in the case of a trihedral angle. For the general case, the formulas on page 39 shows that this sum is $(2n-3)\triangle$, where n is the number of sides of the solid angle.

2 Sicut in figura plana omnes anguli externi, simul sumti, aequales sunt quatuor rectis: ita in corpore solido omnes anguli solidi externi, simul sumti, aequales sunt octo solidis rectis. Per angulum externum intelligo curvaturam seu inclinationem planorum ad invicem, quam metiri oportet ex angulis planis angulum solidum comprehendentibus. Nam illa pars qua aggregatum ex omnibus angulis planis unum angulum solidum facientibus, minus est quam quatuor anguli recti planum facientes, designat angulum externum solidum.

Prop. I   2 As in a plane figure all the exterior angles, taken together, equal four right angles, so in a solid body all the exterior solid angles, taken together, equal eight solid right angles. By exterior [solid] angle I mean the mutual bending or inclination of the planes, which is to be measured with the help of the plane angles which comprise the solid angle. For the part by which the sum of all the plane angles forming a solid angle is less than the four right angles which form a plane, designates the exterior solid angle.

Comments. The proposition of the first sentence was new with Descartes. It is stated by analogy with plane figures (convex polygons) and no doubt Descartes discovered it from reasoning by analogy. For the polygon, the sum of the exterior angles is equal to four right angles $(4\triangle)$, the whole circle; so too for the solid body (convex polyhedron) the sum of the exterior angles is equal to eight right angles $(8\triangle)$, the whole sphere. No proof is given in the manuscript; several proofs are indicated later in this comment.

The proposition of this sentence is important since the proposition or theorem of Paragraph 3 of the manuscript can be derived from it. Since it is not needed or used elsewhere, it can be regarded as a lemma for that proposition or theorem.

This is the first of six propositions in Part I of the manuscript which are singled out as the most noteworthy. It will be referred to as Proposition I.

The second and third sentences relate to exterior solid angles as such. Prouhet translated the Latin "curvaturam" as "courbure" (curvature), but the word "bending" is used instead in the English, as being an appropriate translation and as avoiding some misconceptions.

The second sentence is somewhat obscure, but, once one knows what exterior solid angles are and their properties, there is little doubt as to what was intended (see Section 5, paragraph 9, pages 40–42). The planes referred to are the planes of the exterior solid angle. The "mutual bending or inclination of the planes" refers to the dihedral angles (of this *exterior* angle), which are the supplements of the plane angles of the original solid angle. Thus the latter, taken together, form a means of measuring the exterior solid angle. The third sentence, giving the measure of an exterior solid angle, helps to clarify the second sentence, but some obscurity in terminology results.

The exterior solid angle of a given solid angle is actually not defined or explained: only some things about it are stated. The incompleteness and obscurity strongly suggest that Descartes was dealing with things already known. This was already mentioned in Section 4 (p. 32) in connection with the dating of the original manuscript.

Prouhet gave a demonstration of the proposition of the first sentence, but before describing that, a proof indicated by Pólya[38] will be outlined first. Consider the exterior solid angles constructed at each vertex of the polyhedron in the manner described first in Section 5, paragraph 9 (p. 40). This is analogous to the construction of the exterior angles of a polygon as shown in Fig. 4(b) of Section 5 (p. 38) which is similar to Pólya's Fig. 3.7. The exterior solid angles are closed by describing a portion of a sphere about each, forming each exterior solid angle into a sector of a sphere (as in Fig. 4(b) arcs of circles are drawn forming each exterior angle into a sector of a circle). "The sectors [of the sphere] so generated at the several vertices of the polyhedron form, when shifted together, a full sphere as the circular sectors in the analogous plane figure (Fig. 3.7) form, when shifted together, a full circle." Hence, "the joint measure of all the exterior solid angles of the polyhedron is, in fact, 4π." That the spherical sectors can be shifted to form the full sphere follows from the fact that the plane angles facing each other, of two neighboring exterior solid angles, are congruent and their sides are parallel, hence the two exterior solid angles can be moved without turning to make the two facing plane angles coincide.

Prouhet's proof is based on the construction of an exterior solid angle according to the method described on p. 42, in connection with Fig. 7 of Section 5. It is analogous to the proof for polygons illustrated by Figure 4(c) (p. 38). This analogy is not indicated by Prouhet; our Figs. 3(c) and 4(c) are given in Section 5 only to show the analogy. The proof goes as follows. From a point P in the interior of the polyhedron, drop perpendiculars to each face of the polyhedron. The perpendiculars to the faces of a particular solid angle form the exterior solid angle of that solid angle (as in Fig. 7). Each solid angle of the polyhedron has its corresponding exterior angle with vertex at the point P, and all the exterior solid angles about the point P correspond respectively to the solid angles of the polyhedron. The exterior solid angles, with their common vertex at the point P, cover the space about P without any overlap and hence their sum is equal to the total area of the sphere about this point, 8 solid right angles (4π).

Coxeter[39] gives a different proof, based upon projecting the polyhedron on to the surface of a unit sphere about an interior point as center, but this presumably would not have been thought of in the 17th century.

Descartes may of course have had a proof, but he may not have bothered. The analogy with plane figures is so close and direct that he may have thought further proof unnecessary.

Curiously, the word "analogy" does not appear in the *Méthode* nor in the *"Regulae ad Directionem Ingenii."* In the latter work Descartes admits only two "mental operations by which we are able. . . . to arrive at the knowledge of things"; these are "intuition and induction," the former being quite broadly defined.[40]

3 Si quatuor anguli plani recti ducantur per numerum angulorum solidorum et ex producto tollantur 8 anguli recti plani, remanet aggregatum ex omnibus angulis planis qui in superficie talis corporis solidi existunt.

Prop. 2   *3* If one multiplies four plane right angles by the number of solid angles and from the product one removes 8 plane right angles, the remainder will be the sum of all the plane angles which are in the surface of the solid body.

Comments. The following letters will be used consistently in expressing propositions in this section:

S = number of Solid angles,
F = number of Faces,
P = number of Plane angles,
Σ = Sum of the measures of all the plane angles,
Δ = one right angle = $\pi/2 = 90°$.

Thus, the proposition of this paragraph can be expressed by the formula

$$\Sigma = (4S-8)\Delta. \tag{1}$$

No proof is given by Descartes, but the proposition follows in a simple manner from Proposition 1 and the last sentence of Paragraph 2. This last (third) sentence gives the measure of the exterior solid angle of a given solid angle. For the exterior angle of one of the solid angles of the polyhedron this is four right angles, less the sum of the plane angles of that particular solid angle. There is one such expression for each of the solid angles of the polyhedron; adding them all together produces the result that the sum of all the exterior angles, which by Proposition 1 is equal to 8Δ, is equal to four right angles times the number of solid angles, less the sum of all the plane angles in the polyhedron. Expressed otherwise, $8\Delta = 4S\Delta - \Sigma$, or $\Sigma = (4S - 8)\Delta$.

Note that the above derivation is analogous to the derivation by Proclus of the corresponding proposition for polygons (Section 5, paragraph 5, p. 37).

The proposition of this paragraph is Descartes' basic theorem and will be referred to as Proposition 2. It will be discussed in later sections.

4 In pyramide sunt semper tot facies quot anguli. In columna media pars numeri angulorum solidorum minor est binario quam numerus facierum. In pyramide duplicata media pars numeri facierum minor est binario quam numerus angulorum. Sunt et alia corpora in quibus licet duo extrema imaginari et plures zonas. Sunt ad minimum triplo plures anguli plani quam solidi in uno corpore. Si tollatur binarius ex numero angulorum solidorum qui in corpore aliquo continentur, et residuum ducatur per binarium, fit maximus numerus facierum. Si vero dividatur numerus angulorum per binarium, si quidem sit numerus par, sin minus illi prius addenda erat unitas ut dividi possit, ac postea quotienti addatur binarius, erit numerus minor facierum. Est maxima reciprocatio inter facies et angulos solidos.

4a tetraedron
   cubus
   octaedrum

4 [1] In the pyramid there are always as many faces as solid angles. [2] In the prism, half the number of solid angles is less by two units than the number of faces. [3] In the bipyramid, half the number of faces is less by two units than the number of [solid] angles. [4] And there are other bodies in which one can imagine two extremes and more zones. [5] There are at least three times as many plane angles as solid angles in a body. [6] If one takes two away from the   Prop. 3 number of solid angles contained in a body, and multiplies the remainder by two, one has the maximum number of faces. [7] But if one divides by two the number of solid angles if this number is even—if not, one had first to be added so that it can be divided—and then adds two to the quotient, one will have the minimum number of faces. [8] There is a maximum relation between the faces and the solid angles.

4a tetrahedron
   cube
   octahedron

Comments. The sentences in the English have been numbered for reference.

[1] If the base of the pyramid has n angles and sides, the number of solid angles, including the apex, is $n + 1$, and the number of faces is $n + 1$, the number of lateral faces plus the base. Hence $F = S$.

[2] If the two bases of the prism each have n angles and sides, there are $2n$ solid angles, so $S = 2n$; the number of faces is the sum of the n lateral faces and the two bases, so $F = n+2$. Hence, eliminating n, $S/2 = F-2$, as indicated by the proposition.

[3] A bipyramid is formed by attaching by their bases two pyramids with congruent or similar (congruent by reflection) bases. If the two bases have n angles and sides, the number of solid angles is $n+2$, so $S = n+2$, and the number of faces is $2n$, so $F = 2n$. Hence $S = F/2 + 2$, or $F/2 = S-2$, as stated in the proposition.

The three words of the note, which appears in the margin of the manuscript and is here referred to as Paragraph 4a, are evidently examples to illustrate these sentences. The tetrahedron is an example of a pyramid, the cube of a prism, and the octahedron of a bipyramid.

[4] The preceding three sentences are concerned with types of polyhedra which have a specific relation between the number of faces and the number of solid angles. This sentence is taken as continuing this thought by referring to other types of polyhedra which also have a specific relation between faces and solid angles.

The "two extremes and more zones" can be explained in several ways, as follows. The bipyramid can be characterized as having two extremes, the two apices, with two zones between them, the two belts of triangular lateral faces.

If a prism is interposed between the bases of the two pyramids (in other words, if a pyramid is placed on each of the opposite bases of a prism), we have a solid body with two extremes, as before, and an additional zone, the belt of lateral quadrilateral faces. The relation between faces and solid angles in such bodies is $2F/3 = S-3$. The prism can be characterized as having two extremes, the two bases, with a zone between them, the belt of lateral quadrilateral faces. If the two pyramids forming a bipyramid are truncated, we have a solid body with two extreme faces and two zones of quadrilateral faces between them. Here the relation between faces and solid angles is $F = 2S/3 + 2$. And additional zones or belts of quadrilateral faces can be interposed in either of the above two types, forming still more types in each of which there is a relation between faces and angles. (Note that Kepler[41] in 1619 used the term "zona" meaning "belt" or "zone," for a belt of congruent faces.) The sentence as it reads does not go on to state "in which there is a relation between faces and vertices" or its equivalent; this may be implied from the preceding sentences, or possibly a phrase may have been omitted.

[5] Since each solid angle must have three or more plane angles, the number of plane angles must be at least three times the number of solid angles, so $P \geqq 3S$. At first sight this sentence appears to be irrelevant at this point, but it is useful in connection with one of the sentences which follow.

[6] and [7] The inequalities of these sentences can be written, if one follows the wording and ignores the distinction between odd and even S in the second, as

$$2(S - 2) \geqq F \tag{2a}$$

and

$$S/2 + 2 \leqq F. \tag{2b}$$

Another form is

$$\text{(a) } 2S \geqq F + 4, \quad \text{(b) } 2F \geqq S + 4.$$

The first of these inequalities can be derived from the basic theorem, equation (1), as follows. Each face must have at least three angles and hence the angle sum for a face must be equal to or greater than 2Δ, the angle sum of a triangle. Hence Σ, the total angle sum for the F faces, must be equal to or greater than $2F\Delta : \Sigma \geqq 2F\Delta$. By substituting the value for Σ given by equation (1) we have $(4S - 8)\Delta \geqq 2F\Delta$, or $2S - 4 \geqq F$.

Considering the second inequality in the form $F \geqq (S + 4)/2$, it is evident that it is immaterial whether S is even or odd, for if S is odd, F, being an integer, could only be equal to at least the next integer beyond $(S + 4)/2$: hence it is not necessary to have two equations, one for S even and another for S odd.

The second inequality can also be derived from equation (1). For simplicity, the relation given in Paragraph 11 of the manuscript is used. This is $(4F + \frac{\Sigma}{\Delta})/2 = P$. Since $P \geqq 3S$ (cf. [5] above), we have $4F + \frac{\Sigma}{\Delta} \geqq 6S$.

Substitution of equation (1) for Σ results in $4F + 4S - 8 \geqq 6S$ which reduces to $4F - 8 \geqq 2S$ or $2F - 4 \geqq S$.

These two inequalities will be referred to as Proposition 3.

[8] This sentence must be considered in the light of the preceding sentences of the paragraph. The first three are concerned with types of polyhedra in which there is a specific relation between the number of faces and the number of solid angles and the fourth sentence suggests other types. Apparently Descartes may have been motivated by analogy with polygons, the faces and solid angles of the polyhedron being analogues of the sides and angles of the polygon. But there was no single general relation between faces and solid angles of polyhedra (as there is for sides and angles of polygons), and he could only give different relations for some special types. Following this he gave inequalities which showed on the one hand the maximum number of faces there could be with respect to a given number of solid angles and, on the other hand, what can be said to be the maximum number of solid angles with respect to a given number of faces. The present sentence then would appear to refer to at least the first of the two inequalities.[42]

5 Pyramides omnes aequilaterae in sphaera describuntur.

5 All equilateral pyramids can be inscribed in the sphere.

Comments. An equilateral pyramid is a regular pyramid—the lateral faces are congruent isosceles triangles and the base a regular polygon. This is obviously inscribable in a sphere.[43]

6 Coni rectanguli cuius scilicet altitudo aequatur semidiametro basis, superficies convexa se habet ad basin ut $\sqrt{2}$ ad unitatem, quemadmodum lineae simplices.

6 Of a rectangular cone, namely [a cone] the altitude of which is equal to the semidiameter of the base, the convex surface is to the base as $\sqrt{2}$ is to unity, in the same way as the simple lines.

Comments. A rectangular cone is a right cone in which a vertical cross section through the apex is an isosceles right-angled triangle standing on its hypotenuse; the height is equal to the radius of the base r and the slant height is $\sqrt{2}r$; the ratio of these two lines is $\sqrt{2}r/r \doteq \sqrt{2}/1$. The lateral area is one-half the perimeter of the base ($2\pi r$) times the slant height ($\sqrt{2}r$) or $\sqrt{2}\pi r^2$, and the area of the base is πr^2; the ratio of the two areas is $\sqrt{2}/1$, the same as the above lines.

7 Sic demonstratur non plura esse quam 5 corpora regularia: quia si ponatur α pro numero angulorum solidorum, et $1\mathcal{X}$ pro numero facierum, debet dividi posse

$$\frac{2\alpha - 4}{1\mathcal{X}} \text{ et } \frac{2\mathcal{X} - 4}{1\alpha},$$

ita ut nulla occurrat fractio; alioquin enim certum et evidens est corpus regulare esse non posse. Hoc autem inveniri tantum potest, si α sit

4./.6./8./12./20, et pariter 1⟩ℓ sit 4/8/6/20/12: unde generantur 5 corpora regularia.

Prop. 4   7 As follows it is proven that there cannot be more than 5 regular bodies: since if one takes α for the number of solid angles, and ϕ for the number of faces, one must be able to divide

$$\frac{2\alpha - 4}{1\phi} \quad \text{and} \quad \frac{2\phi - 4}{1\alpha},$$

in a manner that no fraction occurs; otherwise it is certain and evident that a regular body is not possible. But this can only be so if α is 4, 6, 8, 12, 20, and respectively ϕ is 4, 8, 6, 20, 12, by which are generated the 5 regular bodies.

Comments. The cossic sign in the manuscript for the number of faces is replaced by ϕ in the translation, as this appears later in the manuscript (Paragraph 16) for the number of faces.

The two formulas are written

$$\frac{2S - 4}{F} = a, \quad \frac{2F - 4}{S} = b,$$

where a and b are integers. They are derivable from equation (1), page 46, and the properties of regular bodies.

The first one is derivable as follows. All the faces in regular bodies have an equal number of angles, say n; the angle sum for each face is $2(n-2)\Delta$ (Section 5, paragraph 3 p. 36), and for all F faces the total angle sum is $\Sigma = 2(n-2)F\Delta$. Substitution of this value in equation (1) results in

$$2(n - 2)F\Delta = (4S - 8)\Delta,$$

hence

$$\frac{2S-4}{F} = n - 2 = a.$$

For the second equation, all the solid angles have the same number of plane angles, say m, and hence the number of plane angles P is equal to mS. Substitution of this value in equation (3) in Paragraph 11 (below p. 54) gives $\Sigma/\Delta = 2mS - 4F$. Substitution of the value for Σ in equation (1) results in

$$2mS - 4F = 4S - 8$$

or

$$\frac{2F-4}{S} = m-2 = b.$$

It can be shown in several ways that neither a nor b can be greater than 3. For example, the two equations are linear in S and F and can be solved for their values, in terms of a and b. These values are:

$$S = \frac{8 + 4a}{4 - ab}, \quad F = \frac{8 + 4b}{4 - ab}.$$

Hence ab must be less than 4. The possible combinations of integers a and b which give a product less than 4 are (1,1), (1,2), (2,1), (1,3) and (3,1). When these combinations are substituted in the above expressions, the results for S and F respectively are (4,4), (6,8), (8,6), (12,20) and (20,12).

The number of angles (sides) in each face (n above) and the number of plane angles in each solid angle (m above) can thus be determined readily and the five bodies constructed.

A geometric proof that there cannot be more than the five regular bodies is given by the last proposition in Book XIII of Euclid. Descartes' condition is essentially algebraic and may very well have been the first algebraic treatment.

The proposition of this paragraph will be referred to as Proposition 4.

Note that Descartes used the word "faces" for the faces of a solid body, and this is done throughout the manuscript. This usage is unusual and perhaps even unique. The word commonly used was "base," in various languages, and even Euler in 1750 used the Latinized form of a Greek word meaning base. Legendre used "face" in his geometry of 1794. (See Section 8, p. 66.)

8 Rhomboeides omnes et pyramides sphaeram circumscribunt.

8 All rhomboids and pyramids can be circumscribed about the sphere.

Comments. This sentence is too broad as to pyramids, since not all pyramids are circumscribable. Evidently equilateral (regular) pyramids were intended, as in Paragraph 5, or perhaps the word "aequilaterae" was omitted in copying the original manuscript.

It is not evident what Descartes meant by the term rhomboid. If he was acquainted with Kepler's *Harmonice Mundi* (of 1619), it may refer to the two rhomboids there described and illustrated, the rhombic dodecahedron having 12 rhombic faces and the rhombic triacontahedron having 30 rhombic faces, to which the rhombohedron with 6 rhombic faces could be added. (All three were described in an earlier work of Kepler, see Section 12, p. 119.) Each of these is circumscribable about a sphere.

Prouhet II (ignoring Kepler) suggested that the word referred to a "regular double pyramid" (regular bipyramid) as an extension of the term as used by Archimedes. In his "On the Sphere and Cylinder"[44] Prop. 18, Archimedes referred to two right cones joined by their congruent circular bases as a solid rhombus (rhomboid). The cones did not necessarily need to have equal altitudes and his use of the term was actually an extension of an older use. According to Heath[45] two congruent right cones joined by their bases may have been once called rhomboids since the section by every plane through the axis is a rhombus, a plane figure defined by Euclid as one having four equal sides but not right angles, I def. 22. Prouhet's suggestion, then, is that Descartes extended the term to the bipyramid formed by the union of two regular pyramids; any of these (not necessarily formed from congruent pyramids as Prouhet supposed) would obviously be circumscribable about a sphere.

9 Ut cognoscamus utrum aliquod corpus solidum possit in sphaera describi, primo sciendum est omnes eius facies necessario in circulo describi posse. Quo posito, si tres anguli unius faciei aequaliter distent a centro sphaerae, certum erit etiam alios omnes eiusdem faciei aequaliter a centro sphaerae distare; ac insuper ex consequenti, angulos omnes vicinarum facierum, qui simul concurrunt cum illis prioris faciei in iisdem angulis solidis.

9 To know if any solid body can be inscribed in a sphere, it is first necessary to know that all its faces necessarily can be inscribed in a circle. That given, if three angles of one face are equally distant from the center of the sphere, it is certain that all the others of the same face will also be equally distant from the center of the sphere; and moreover, consequently, all the angles of the adjacent faces, which meet with those of the first face in the same solid angles.

Comments. A necessary condition that a polyhedron be inscribable in a sphere is, obviously, that all the faces be inscribable in circles. However, this condition is not sufficient (it is for pyramids, but not in general). A simple counterexample is the bipyramid; each face, being a triangle, is inscribable in a circle, but the solid itself is not inscribable in a sphere unless two other conditions are also present.

This paragraph appears to be asserting that the given necessary condition is also sufficient and attempts the beginning of a demonstration. It has the appearance of something which the author started, but then, realising that it was not going right, laid aside to come back to at a later time. The last statement in the paragraph, beginning "and moreover, consequently,..." does not follow from and is not a consequence of the preceding statement.

It is not surprising that Descartes could not give the necessary and sufficient conditions for any convex polyhedron to be inscribable in a sphere, for these are not known even today.[46] Perhaps he generalized from pyramids and prisms, and the regular and semiregular polyhedra, in which special cases inscribability of each face is sufficient, and assumed that the only thing that was necessary, and therefore sufficient, was that each face be inscribable in a circle (if so, induction failed again). This is in fact true for trilinear (simple) polyhedra, in which each vertex has only three lines and three faces, but it may be doubted that he had only these in mind here as he refers to *"any* solid body" and he was aware of inscribable nontrilinear polyhedra (see Section 11, p. 98ff.).

The theorem that a trilinear convex polyhedron is inscribable in a sphere if and only if each of its faces is inscribable in a circle is implied by Exercise 2 in Grünbaum, *Convex Polytopes.*[47] Such a proposition could have been stated and proved in Descartes' time or even in the time of Euclid, except that apparently no one thought of dealing separately with such a class of polyhedra until the 19th century (but Descartes did consider solid bodies with all faces triangles, see Paragraph 16 below). A simple proof of this special case, which parallels the attempted proof of the general case, follows.

Figure 8 represents a portion of a convex polyhedron in which only three faces meet at each vertex and every face is inscribable in a circle. Consider face 1 and its adjoining faces 2, 3, 4,..., n. A sphere (and only one sphere) can be drawn through the four points A, B, C, F, since they are not coplanar. This sphere contains the vertices A, B, C of face 1 and hence contains the circumscribing circle C_1 of face 1 (only one circle can be drawn through three points), and therefore it contains all the vertices of face 1, including vertex D. Similarly, this same sphere, containing vertices A, B, F of face 2, contains the circumscribing circle of face 2 and all its vertices. This same sphere contains vertices B, C, F of face 3 and hence contains all the vertices of face 3, including vertex G. Vertices C, D, G of faces 1 and 3 are on this sphere, but these are also vertices of face 4 and hence all the vertices of face 4 are on the sphere. The same reasoning applies to the consecutive remaining faces adjoining face 1. Hence, the vertices of any face of the trilinear polyhedron and all the vertices of each face adjoining it lie on a single sphere.

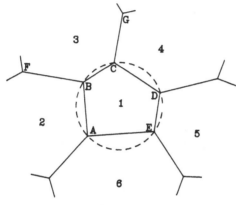

Figure 8

10 Dato aggregato ex omnibus angulis planis qui in superficie alicuius corporis solidi existunt, invenire quot in eodem corpore solidi anguli existant. Addantur 8 numero dato, et productum dividatur per 4: residuum erit numerus quaesitus, ubi si fractio ocurrat, certum est nullum tale corpus esse posse.

10 Given the sum of all the plane angles which are on the surface of a solid body, find how many angles there are in that solid body. Add 8 to the given number and divide the result by 4; the result will be the number sought, but if a fraction occurs, it is certain that such a body cannot exist.

Comments. The rule given by the second sentence can be expressed by the formula $(\Sigma/\triangle + 8)/4 = S$. This is simply Proposition 2 (equation (1), page 46) solved for S.

11 Dato aggregato ex omnibus angulis planis et numero facierum, numerum angulorum planorum invenire. Ducatur numerus facierum per 4, et productum addatur aggregato ex omnibus angulis planis: et totius media pars erit numerus angulorum planorum. V.g., aggregatum ex omnibus angulis planis est 72, numerus facierum 12, cuius quadruplum 48 additum cum 72 facit 120, cuius media pars est 60: ergo in tali corpore sunt 60 anguli plani.

11 Given the sum of all the plane angles and the number of faces, find the
Prop. 5 number of plane angles. Multiply the number of faces by 4, and add the
product to the sum of all the plane angles: and the half of all will be the number
of plane angles. For example, if the sum of all the plane angles is 72 [right
angles], and the number of faces 12, the quadruple of the latter, 48, added to 72
gives 120, the half of which is 60; hence in such a body there are 60 plane angles.

Comments. The rule given by the second sentence, as stated, can be
expressed by the formula

$$(4F + \tfrac{\Sigma}{\Delta})/2 = P. \tag{3}$$

Arranged differently, this gives us a second angle sum formula;

$$\Sigma = (2P - 4F)\Delta. \tag{3a}$$

This formula can be derived simply by adding together the individual
formulas for the sum of the interior angles of each polygonal face. If n is the
number of plane angles (sides) of one face, the interior angle sum is $2(n-2)\Delta$
or $2n\Delta - 4\Delta$ (Section 5, paragraph 3, p. 36). There are F such expressions, and
adding them together produces the result that the total sum of all the plane
angles, Σ, is equal to 2Δ times the sum of the n's, or $2P\Delta$ (P being the total number of plane angles), less 4Δ times the number of faces F: so $\Sigma = (2P - 4F)\Delta$.

For the given example, which has 72Δ for the sum of all the plane angles,
equation (1) shows that there are 20 solid angles. Very probably the regular
dodecahedron, with 12 faces and 20 solid angles, was used to make up the
example. There are, however, 7595 combinatorially distinct dodecahedra
which satisfy the given and derived conditions.[48]

The proposition of the second sentence, equation (3), will be referred to as
Proposition 5.

12 Sunt semper duplo plures anguli plani in superficie corporis solidi
quam latera; unum enim latus semper commune est duobus faciebus.

12 There are always twice as many plane angles as sides on the surface of a
solid body; for one side is always common to two faces.

Comments. This statement is not referred to elsewhere nor is it used or
needed for anything in this part of the manuscript. A side of a face is still the
side of a face even though two sides coincide where two faces have a side
in common.

The statement comes from Pappus and is relevant to Part II of the
manuscript: see Section 11 (n.113), where further reference is made to it.

13 Si omnes facies dicantur aequalem numerum angulorum planorum
continere, ergo numerus angulorum dividi poterit per numerum facierum sine
fractione, et quotiens erit numerus angulorum unius faciei. Hinc facile
cognoscetur, ex numero angulorum planorum et numero facierum solum
cognitis, quot anguli in una facie esse debeant. V.g., si sint 5 facies et 18 anguli
plani, ergo ex illis faciebus vel 2 erunt triangulares et 3 quadratae, vel 3
triangulares una quadrata et altera pentagona, vel denique una hexagona et 4

triangulares. Sed quia in eodem corpore sunt 6 anguli solidi, hinc non potest ullum tale corpus existere, nisi cuius sint... .

13 If all the faces are said to contain an equal number of plane angles, then the number of angles can be divided by the number of faces without fraction, and the quotient will be the number of angles of one face. Hence one will know easily, from knowing only the number of plane angles and the number of faces, how many angles there must be in one face. For example, if there are 5 faces and 18 plane angles, then, of these faces, either 2 will be triangular and 3 quadrilateral, or 3 triangular, one quadrilateral and the other pentagonal, or finally one hexagonal and 4 triangular. But since there must be 6 solid angles in this body, no such body can exist unless it has *2 triangular faces and 3 quadrilateral.*

Comments. The three dots indicating an omission are in the manuscript; the supplement in italics in the translation at this point follows a proposal by Prouhet and de Jonquières.

The example poses a polyhedron with 5 faces and 18 plane angles. Since each face must contain at least 3 plane angles, 15 are accounted for, leaving 3 which must be distributed among the 5 faces. This can be done in only three ways: one given to each of 3 faces; 2 given to one face and one to another; and all 3 given to one face. This results in the three combinations of types of faces given in the text:

(a)   3, 3, 4, 4, 4,
(b)   3, 3, 3, 4, 5,
(c)   3, 3, 3, 3, 6.

The total angle sums for each of these combinations can easily be calculated from the formula $2(n-2)\Delta$ for a face, and the result is 16Δ for each combination. But, even simpler, substitution of $F=5$ and $P=18$ in equation (3a) (p. 54) gives 16Δ for the total angle sum. Substitution of this value for Σ in equation (1) (p. 46), gives $S=6$. Hence any polyhedron satisfying the given conditions must have 6 solid angles.

It must still be shown that such a polyhedron is possible. Combinations (b) and (c) are obviously impossible. Combination (a) is satisfied only by the prism with a triangular base; hence such a polyhedron exists.[49]

14 Triplicem adverto in angulis solidis aequalitatem aut inaequalitatem: aequales dicuntur qui aequali numero angulorum planorum comprehenduntur; aequales item qui aequalem inclinationem continent, quo casu dicemus angulos externos sive inclinationis aequales esse, et priores dicemus aequales arithmetice; ac denique maxime proprie aequales dicuntur, qui eandem partem sphaerae comprehendunt, et dicentur capacitate aequales.

14 I note a triple equality or inequality in solid angles: those comprised under an equal number of plane angles are said to be equal; also equal are those which have an equal inclination, in which case we will say the exterior angles or inclinations are equal, and the first we will say are equal arithmetically; and finally most properly said to be equal are those which intercept the same part of the sphere, and we will say that they are equal in capacity.

Comments. The three types of equality may be stated as follows, with respect to the spherical polygons intercepted on the sphere:

1. Equal number of sides,
2. equal perimeters,
3. equal areas.

In the third type of equality, the most proper one, the solid angles have the same measure. As to the second type, the text states that the solid angles "have an equal inclination." Descartes refers to the sum of the plane angles of a solid angle (the perimeter of the spherical polygon) as the inclination of the solid angle (see Section 5, paragraph 8, p. 40). If the inclinations of two solid angles are equal, the exterior angles are equal in measure, since the measure of an exterior solid angle is $4\triangle$ less the sum of the plane angles (the inclination) of the given solid angle. The next phrase, reading "in which case we will say the exterior angles or inclinations *are equal*" (the last two words supplemented), refers to this relation.

There is, of course, a fourth type of equality, in which the solid angles are congruent, directly or by reflection.

15 Angulorum solidorum inclinatione aequalium ille capacitate maior est, qui arithmetice exuperat; et omnium capacissimus est angulus coni.

15 Of solid angles which have the same inclination, the largest in capacity is the one which exceeds arithmetically; and the largest of all in capacity is the angle of the cone.

Comments. In terms of the polygons intercepted on the sphere by the solid angles, the first part of this statement would read: Of spherical polygons which have equal perimeters, the one with the largest number of sides has the largest area.

The statement in the text is obviously too broad, but the intention is clear. Prouhet states that Descartes undoubtedly meant regular solid angles (that is, those which have equal face angles and equal angles of inclination of the planes) and that the proposition amounts to "Of two regular and isoperimetric spherical polygons, the one with the largest number of sides has the largest area." He does not comment on the cone, which must be a circular cone which intercepts an isoperimetric circle on the sphere.

The proposition of this paragraph is analogous to the corresponding theorem for plane regular polygons and the circle, known to the Greeks and developed by Pappus,[50] but it is incompletely expressed.

16 Ponam semper pro numero angulorum solidorum α et pro numero facierum ϕ. Aggregatum ex omnibus angulis planis est $4\alpha-8$, et numerus ϕ est $2\alpha-4$, si numerentur tot facies quot possunt esse triangula. Numerus item angulorum planorum est $6\alpha-12$, numerando scilicet unum angulum pro tertia parte duorum rectorum. Nunc si ponam 3α pro tribus angulis planis qui ad minimum requiruntur ut componant unum angulum angulorum solidorum, supersunt $3\alpha-12$, quae summa addi debet singulis angulis solidis iuxta tenorem quaestionis, ita ut aequaliter omni ex parte diffundantur. Numerus verorum angulorum planorum est $2\phi+2\alpha-4$, qui non debet esse maior quam $6\alpha-12$; sed si minor est, excessus erit $+4\alpha-8-2\phi$.

16 I always take α for the number of solid angles and ϕ for the number of faces. The sum of all the plane angles is equal to $4\alpha-8$, and the number ϕ is equal to $2\alpha-4$ if one counts as many faces as there could be triangles. The number of plane angles is $6\alpha-12$, by counting a [plane] angle for a third of two right angles. Now if I take 3α for the three plane angles which are the least number required to form one of the solid angles, there remains $3\alpha-12$, which sum must be added to the individual solid angles according to the investigation, so that they will be distributed equally from every part. The actual number of plane angles is $2\phi+2\alpha-4$, which cannot exceed $6\alpha-12$, but if it is less, the excess will be $+4\alpha-8-2\phi$.

Prop. 6

Comments. In what follows, S is used for the number of solid angles instead of α and F instead of ϕ for the number of faces.

The first part of the second sentence repeats the proposition of equation (1):

$$\frac{\Sigma}{\triangle} = (4S-8). \tag{1}$$

The second part of the second sentence indicates that if all the faces are triangles,

$$F = 2S-4.$$

This can be derived as follows. Since the angle sum for each triangle is $2\triangle$, the total angle sum is $2F\triangle$, and substitution of this value for Σ in (1) results in $F = 2S-4$.

The third sentence continues the case of polyhedra with all faces triangular. The number of plane angles is

$$P = 6S - 12.$$

Each triangle, of angle sum $2\triangle$, has three plane angles, hence the average of the measures of all the plane angles is $2\triangle/3$. The number of plane angles P is then given by Σ, the total angle sum, divided by $2\triangle/3$, hence $P=3\Sigma/2\triangle$. $\Sigma/\triangle$ is therefore $2P/3$, and substitution of this value in (1) results in $P = 6S - 12$.

The fourth sentence also assumes that all faces are triangles. The total number of plane angles is $6S-12$. Each solid angle must have at least 3 plane angles; this accounts for $3S$ plane angles, and the balance, $3S-12$, must be distributed among all the solid angles. But it is only in the case of the regular octahedron and the regular icosahedron, which have all faces triangles, that the excess of plane angles (above three per solid angle) can be distributed equally among the solid angles.

The last sentence is general and refers to the fact that $6S-12$ is the maximum number of plane angles (with respect to a given S). This is evident from the preceding paragraphs; the smallest possible value of the averages of all the plane angles is $2\Delta/3$, hence dividing Σ by this smallest value gives the largest possible value for the number of plane angles with respect to a given S.

We repeat the last sentence, but using the symbols F and S in the equations: "The actual number of plane angles is $2F + 2S-4$, which cannot exceed $6S-12$, but if it is less, the excess will be $4S-8-2F$." This sentence, which is derived simply by combining equations (1) and (3), is true for any convex polyhedron, and is formulated as

$$P = 2F + 2S-4. \tag{4}$$

Descartes stops here; there is no discussion of this formula nor is anything done with it except to derive the expression of the deficiency of the number of plane angles from the maximum number possible.

The proposition of the last sentence will be referred to as Proposition 6 and will be considered in later sections.

17 Describi possunt Rhomboeides in sphaera cuiuscumque quantitatis, sed non aequilatera.

17 Rhomboids of any quantity, but not equilateral, can be inscribed in the sphere.

Comments. No satisfactory meaning can be derived from this sentence, nor has any suitable revision been suggested.[51] Any attempt to make out a meaning is frustrated by the expression "but not equilateral." But even if this phrase were removed or changed to "but not unless equilateral," difficulties would still remain. The Kepler rhomboids referred to under Paragraph 8 are not inscribable, nor are the bipyramids mentioned under Paragraph 8 inscribable, unless there is a special relation between the altitudes of the two pyramids.

7 General Comments

The preceding section discussed the individual paragraphs of the manuscript, primarily from the standpoint of explanation and derivation. The present section offers some general comments and observations.

At first reading the manuscript gives the impression of a miscellaneous set of disorganized notes; many have no relation to any of the others and for some the relation to others is not immediately apparent. But some of the paragraphs, if separated from the others, do form a connected, interdependent whole in substantially good order. These are Paragraphs 1–4, 7, 10, 11, 13, 16. The remaining paragraphs, 5, 6, 8, 9, 12, 14, 15, 17, are those which introduce the disorder. None of the statements in this second group is referred to in the first group nor are any of them useful for the derivation of any of the statements of the first group. And, except that two of them deal with solid angles, none of them refers to or utilizes any of the statements of the first group.

The second group is heterogeneous. Paragraph 5 relates to inscribed pyramids, Paragraph 17 to inscribed rhomboids, and Paragraph 9 attempts a general statement on inscribability of polyhedra. Paragraph 6 gives the ratio of the lateral area of a cone to the area of the base. Paragraph 8 refers to certain circumscribable polyhedra. Paragraph 12 is simply an isolated observation, perhaps relevant to Part II of the manuscript. Paragraph 14 is related to Paragraph 15 in that its definitions are needed for understanding 15, which, indirectly, draws an analogy between the isoperimetric problem of plane polygons and spherical polygons.

Excluding Paragraph 12 (which will be mentioned later), several things should be noted with respect to this second group. First, they are mainly geometric in character, with no algebra involved except in calculating the area of a cone. Second, they contain most of the incorrect statements of the manuscript, some of which can be made correct by suitable amplification or modification, but others not; of course, some of these may be due to faulty copying rather than to the original manuscript. Third, they do not depart in character from the ordinary geometry of the time, and if new do not add anything significant. (The same holds true for Paragraph 9, since the attempted geometric proof failed.)

On the other hand, the first group of paragraphs is homogeneous, interdependent, and substantially well ordered; there is only one incorrect statement (of no consequence), once the statements are understood; after Proposition 1 all propositions and corollaries are derivable by simple algebra; and the subject matter represents a considerable departure from contemporary thinking about polyhedra. The last two statements will be amplified.

Concerning polyhedra, there was very little known in the early 17th century beyond what had been transmitted by the ancient Greeks.[52] Panofsky states that the geometry of three-dimensional bodies or stereometry "was entirely disregarded during the Middle Ages."[53] Works up to the time of Descartes dealt with descriptions, metrical properties, problems of mensuration, and the generation of new bodies. In about 1475 the artist Pietro Franceschi wrote "De corporibus regularibus,"[54] in which he treated the five regular bodies from the standpoint of inscribability in and circumscribability about each other and the circle, and the calculation of various relations and quantities; a typical problem is to find the area of a cube circumscribed about a sphere of diameter 7. The last section, on "irregular bodies," has two of the semiregular (Archimedean) bodies and a few others. The first section is plane geometry and treats regular polygons; a sample problem is to find the area of a regular octagon circumscribed by a circle of diameter 7. This is a very dull work, consisting mainly of a series of numerical problems and the working out of their solutions. In 1497 Luca Pacioli showed perspective drawings of the regular solids, several of the Archimedean solids, and some truncated solids (that is, with a pyramidal piece cut off each vertex) and augmented solids (with pyramidal pieces added to each face), and gave brief descriptions.[55] Dürer's book on practical geometry, of 1525, which went through a number of editions including a Latin edition in 1605, had a section (Book 4) on solid bodies.[56] He described the regular solids, most of the Archimedean solids, and several of his own invention. He represented the solids by developing them on a plane sheet in such a way that the faces were connected and the sheet could be cut in one piece and folded to form a model of the solid, a method now commonly used and evidently original with him. The famous engraving Melencolia I of 1514 shows a stone block in the shape of a truncated rhombohedron (the two corners farthest apart truncated); this may have been a departure from previous truncations in that the body to be truncated was not a regular one and only two of the eight corners were truncated. In 1568 Jamitzer published an extraordinary set of 120 perspective drawings of the five regular solids and other solids derived from them by truncations, cutting notches into sides, making concavities in faces, adding pieces to faces, and combinations of these operations, all done in a regular manner.[57] Only a few of the Archimedean solids are included, and one of the drawings appears to anticipate one of Kepler's star polyhedra. Kepler's *Harmonice Mundi* of 1619 has a section (Book II) on plane and solid figures, mainly the latter. He described and illustrated with perspective drawings all the regular and semiregular bodies, the rhomboids, and two stellated (augmented) or star polyhedra, semibodies, as he called them; he was concerned with how various solids were derived from others and with inscribability and circumscribability and metrical relations. Of works written later than the Descartes manuscript mention should be made of those by Ozanam in 1691 and Sharp in 1717.[58] Ozanam's mathematical dictionary (actually a summary arranged by subjects) adds little; he does use the word polyhedron (polyèdre) but strangely limits it to inscribable bodies. Part 2 of Sharp's *Geometry Improv'd* is entitled "A Concise Treatise of Polyhedra or Solid Bodies of Many Bases," but the work is devoted to the construction and dimensions of 12 new bodies. His expression "Polyhedra or Solid Bodies of Many Bases" indicates the etymology of the word and he uses the word "base" throughout for the faces of the solid body. He begins his treatment by stating that the discussion of solid bodies is a neglected and not common part of Geometry.

The above indicates the "state of the art" at the time of the manuscript. The originality and freshness of Descartes' approach are apparent.

In the analysis in Section 6 various propositions were singled out for further mention and numbered. These are listed here:

Proposition 1. The sum of the exterior solid angles of a solid body is equal to eight solid right angles. (§2, p. 44)

Proposition 2. The sum of the plane angles of the faces of a solid body is equal to 4 times the number of solid angles minus 8. By formula, $\Sigma / \Delta = 4S - 8$. (§3, p. 46)

Proposition 3. Two inequalities relating the number of faces and the number of solid angles. By formulas, $S \geq F/2 + 2$ and $F \geq S/2 + 2$. (§4, p. 47)

Proposition 4. Algebraic treatment of the number of regular solid bodies. (§7, p. 50)

Proposition 5. Expression connecting the number of plane angles, faces, and the sum of the plane angles. By formula, $P = (4F + \frac{1}{\Delta})/2$ or $\Sigma = (2P - 4F)\Delta$. (§11, p. 54)

Proposition 6. Formula connecting the number of plane angles, faces and solid angles, $P = 2F + 2S - 4$. (§16, p. 57)

Two characteristics in this group of paragraphs should be noted. The first is their essentially algebraic nature (after Paragraph 2). Paragraph 1 is a geometric definition, and Paragraph 2 (Prop. 1), for which no proof is given, seems to be provable only by geometric means. Thereafter the Propositions and their stated corollaries and applications are derivable in a simple algebraic manner; each follows from other statements in this group, and this can be done without going outside the manuscript itself except for a few known things (such as the known expressions for the sums of the interior angles of a polygon, and the definition of regular solids). Of course the derivations given here in Section 6 may or may not have been those used by Descartes; but there is no suggestion of geometric proofs.

The second characteristic of this group of paragraphs is that they are based on analogy with plane figures. Proposition 1, from which all the other propositions and corollaries flow, is presented as an analogy with the corresponding proposition for polygons and was obviously suggested or discovered by this analogy. And the derivation of some of the statements merely follows a corresponding derivation for plane figures. There are no signs of induction (induction in the sense used by Pólya, referred to below, p. 67) except possibly in Paragraph 4, where the first four sentences may have resulted from an attempt to discover by induction a relation between solid angles and faces, by analogy with the simple relationship which holds between angles and sides in plane figures.

However, this treatise of Descartes did not, apparently, have any bearing or influence on the development of mathematics in the 17th and 18th centuries. Descartes himself apparently did not return to this particular subject. Leibniz, who had a copy of the manuscript, did not do anything with it. Others may have seen the manuscript while it was in the hands of Clerselier, but there is no evidence that any one of those who did see it was influenced by it. Nevertheless, there is a good deal of intrinsic interest in the treatise, and it belongs to the history of mathematicians if not to the history of mathematics as such. It is a work of Descartes which illustrates the development of his thinking: it displays considerable originality, and departs from the manner in which polyhedra had theretofore been treated. It is in fact the first attempt at a general theory of polyhedra. Moreover, it still had something new to offer when it came to light (see below, p. 63).

One can only speculate why Descartes did not return to the preliminary notes and develop them into a memoir for publication. De Jonquières suggests that it was "perhaps because, absorbed in questions of a different order, he never found the time to return to this mathematical production the importance of which, as with that of all other similar ones, was, as he himself wrote, but secondary in his eyes." Foucher de Careil, in his lengthy introduction on the *Méthode,* considers the various manuscripts which he had discovered to be applications of the *Méthode.* This suggests speculation along a different line. When one also takes into account the fact that these notes were written down about 1630, one may conjecture that Descartes contemplated them as the beginning of a mathematical example to illustrate his *Méthode,* and then, realising that the results could not be carried much farther (the consequences of Proposition 1 having been carried out about as far as they could be at the time), or not considering them sufficiently important, he turned to another path, which led to the *Géométrie* with its initiation of analytic geometry and important innovations in algebra. The primarily algebraic nature of the notes even suggests that they may have been part of the beginnings of the thinking which led to the algebraization of geometry. If this admittedly speculative suggestion has any truth in it, then the failure of the notes on polyhedra led to results which were considerably more important.

Once the copy of the manuscript was discovered and published in 1860 it attracted attention. The first to comment on it was Prouhet in his note (Prouhet 1) of April 23, 1860. He stated that the manuscript was worthy of attention from the historical point of view, as an illustration of the thinking of Descartes on the theory of solids, and from the scientific point of view as containing an important new proposition. This is what has here been called Proposition 1; the sum of the exterior solid angles (the polar or supplementary angles) of a solid body (convex polyhedron) is equal to eight solid right angles (4π). He gave the Latin text and a French translation, and remarked that the analogy between polygons and polyhedra had not been noticed before. The proposition was new in 1860. Prouhet called it "Descartes' Theorem" (which is the name it should retain,[59] and showed how it could be derived (his demonstration is reviewed above, Section 6, p. 45). Prouhet considered it obvious that Descartes was dealing only with convex polygons and convex polyhedra.

In his full treatment, later in the same year (Prouhet II), Prouhet again emphasized the importance of the manuscript because of "a very beautiful and general theorem which ought to be placed at the head of the theory of polyhedra...," again referring to Proposition 1, Descartes' Theorem. Prouhet thus put his finger on what was actually still new in his day and what was even then of some importance as a contribution (but his suggestion that it replace Euler's Theorem as the foundation of the theory of polyhedra did not come about). He does not, however, state that the manuscript discloses Euler's Theorem and though he shows how the latter can be derived from Descartes formulations, he does not intimate that Descartes was aware of Euler's Theorem.

Prouhet (II) repeated the demonstration of Proposition 1 which he had given in his first note, but added that it was probably not that of Descartes "who appeared to have been guided here by considerations very important in other respects." From the use of the word "curvaturam" in the manuscript (which he translates as "courbure" or curvature), he conjectures that Descartes proceeded from considerations of the total curvature of a curved surface. But it is not clear how this could have been done. By the curvature of a curved surface is meant the Gaussian curvature; the total curvature of any closed convex surface is 4π ($8\triangle$), the total curvature of a sphere. The relationship between the proposition of Descartes and this total curvature had been pointed out by J. Bertrand in a note immediately following that of Prouhet (I) in the *Comptes rendus*.[60] Prouhet refers to Bertrand, but the latter had also said: "While making this comparison, which comes naturally to mind, one must add however that the beautiful conception of Gauss could not in any manner be considered as a corollary of that of Descartes," and discounted the significance.

Prouhet, following his comment on the first theorem, also states that the first traces of topology (then referred to in French as "géométrie de situation") are found in the Descartes manuscript. He referred to a remark made two years before by L. Poinsot, [61] that what made the theory of polyhedra very difficult was that it bore on a new science, the "géométrie de situation," for it dealt not with the size or proportion of figures but with the order and [relative] position of their elements. However, it does not seem that Prouhet either stated or believed that Descartes had any notion of topology.[62]

Descartes' Theorem is actually a limiting or special case of the Gauss-Bonnet theorem which in its simplest form is that the integral of the Gaussian curvature over a closed surface of genus 0 (one that can be deformed into a sphere) is equal to 4π.[63]

The first formula for the sum of the plane angles (Proposition 2) was known at the time of the publication of the manuscript but the other (Proposition 5), which includes the number of plane angles, was not, nor was the second formula involving the number of plane angles (Proposition 6) known. However, these were not considered significant, and they received little attention. Becker used them in 1869 and Lalanne in 1872 to develop other relations.[64] Other references to them will be noted in Section 9. The algebraic treatment of the number of regular polyhedra does not appear to have been utilized or developed further. The one commonly given in textbooks appeared in 1813 (Lhuilier, see below, p. 70), if not earlier.

By far the greatest number of references to the Descartes manuscript, direct or indirect, occur in connection with its relation to the later work of Euler. We will treat the relationship to Euler below in Section 9, after first reviewing Euler's work on the topic in Section 8.

8 Note on the Euler Papers of 1750 and 1751

We next give a brief review of Euler's work on the general theory of polyhedra. Two papers are relevant here: the first, giving some general results and stating his theorem, was read on November 25, 1750, and the second, giving proofs, was read on September 9, 1751. Both were included in the proceedings of the St. Petersburg Academy for the year 1752–1753, which was published in 1758.[65] The first paper was preceded by a few weeks by a letter to Goldbach summarizing some of the results it contained.[66]

Euler was the first to have any notion of topology. He gave it a name, "geometria situs," which in French was rendered as "géométrie de situation" or "géométrie de position" and in German as "Geometrie der Lage." He derived the name from a statement of Leibniz which, however, may have been misunderstood or garbled in transmission, for it had nothing to do with topology and in fact related to its antithesis; Freudenthal refers to Euler's "terminological mistake."[67] Despite this, his work on polyhedra is presented as a study in solid geometry, stereometry (the measure of solids) as it was then called; perhaps this was because solid geometry was an established subject. Even so, most of the results presented were topological in nature and his proof of the polyhedron theorem was also topological.

The object of the first paper was a general study of polyhedra; one thing which needed to be done was to classify them all in some manner and introduce some order in the mass of diverse solids. Euler quickly abandoned analogy with plane figures, as inadequate. In the case of plane figures (polygons) the matter is simple: the number of sides is equal to the number of angles, and polygons are classified according to the number of sides. But this will not do for solids where, for two different bodies, the number of faces (the two-dimensional analogue of the one-dimensional side of a polygon) can be the same while the number of solid angles (the three-dimensional analogue of the two-dimensional plane angle) is different. Thus a pyramid with a quadrilateral base and a triangular prism each have five faces, but the first has five solid angles and the other has six. (The above is a summary of Euler's own argument; one can conjecture that he may have reasoned further that neither will the solid angles alone do, nor the faces and solid angles taken together.)

Analogy failing, Euler saw the need for something in addition to the two elements corresponding to plane figures, and introduced a concept for which there was no counterpart in the plane. This was the "edge," the ridge formed by the meeting of two adjacent faces along their common side. Since it was new he could find no term in current use, and so selected the Latin word "acies," which has the meaning of a sharp edge. (Legendre, in his *Éléments de Géométrie* of 1794 translated this into French as "arête," which has the meaning of ridge, the sharp crest of a mountain; English has borrowed the French word for this meaning.) Pólya states, "Euler was the first to introduce the concept of the 'edge' of a polyhedron and to give a name to it (acies). He emphasizes this fact, mentioning it twice."[68] Euler concluded that solid bodies had to be studied by considering all their boundary elements, which he defined. Several paragraphs are quoted in translation.

> 5. The consideration of solid bodies therefore must be directed to their boundary; for when the boundary which encloses a solid body on all sides is known, that solid is known, in the same manner that the essential nature of a plane figure customarily is defined by its perimeter.
>
> 6. But to the boundary of every solid body enclosed by plane figures belong:
>
> First: the same plane figures which constitute its boundary, which are named *faces* (hedrae);
>
> Second: the meeting of two faces along their sides, which form the linear bounds of the solid: as I do not find any special name in the writers on stereometry I shall call them *edges* (acies);
>
> Third: the points in which three or more faces meet, which points are called *solid angles*.
>
> 7. Therefore three kinds of bounds are to be considered in any solid body; namely 1) points, 2) lines, 3) surfaces, or, with the names specially used for this purpose: 1) solid angles, 2) edges and 3) faces. These three kinds of bounds completely determine the solid. But a plane figure has only two kinds of bounds which determine it, namely 1) points or angles, 2) lines or sides.[69]

The edge has already been mentioned. Euler does not use the word "face" (Latin facies) but uses "hedra," instead. This is the Latinized form of the Greek word ἕδρα, meaning base, which is the root of the word polyhedron, "with many bases." Note that Sharp in 1717 referred to polyhedra as solid bodies of many bases; Pacioli in 1497 also used "bases" (basi) for faces. As to solid angles, Euler retained the term but shifted its significance to the point of the solid angle; there was no separate term for the point (tip), as such, though "vertex" was used for the apex of a pyramid. (Legendre continued to use "solid angle" in 1794, but Cauchy used "sommet" in his "Recherches sur les polyèdres" of 1813.)

Lakatos emphasizes Euler's innovations and their topological nature as follows:[70]

> The key to Euler's result was just the invention of the concepts of *vertex* and *edge:* it was he who first pointed out that besides the number of faces the number of *points*

and *lines* on the surface of the polyhedron determines its (topological) character. It is interesting that on the one hand he was eager to stress the novelty of his conceptual framework, and that he had to invent the term *'acies'* (edge) instead of the old *'latus'* (side), since *latus* was a polygonal concept while he wanted a polyhedral one, on the other hand he still retained the term *'angulus solidus'* (solid angle) for his point-like vertices.

Euler's main theorem, his Proposition IV, is stated as follows:

In every solid body bounded by plane faces the sum of the number of solid angles and the number of faces exceeds the number of edges by two.

Then, taking the number of solid angles as S, the number of faces as H, and the number of edges as A, the theorem is expressed by the formula:

$$S + H = A + 2.$$

Euler then states that he must confess that he was not yet able to give a definite proof, but that its truth will be recognized for all types of solids for which it is considered. The theorem is then demonstrated as true for: (1) all pyramids; (2) wedge-shaped bodies, that is pyramid type bodies having a line instead of a point for apex; (3) all prisms; (4) prism type bodies with the two bases having different numbers of sides (prismoids); (5), (6) and (7) combinations of two of type (4) joined base to base; and (8), the five regular solids treated individually. The truth of the general proposition is left to be inferred from the individual cases.

Since analogy had proved fruitless, it is very probable that Euler discovered his theorem by induction. Pólya repeatedly uses this as a classic example of induction in mathematical discovery.[71] Euler had come to the conclusion that the three particular types of element had to be considered in characterizing polyhedra. Their values are considered for a number of different specific polyhedra, and a certain type of regularity is then tested for additional cases, and for still more cases, until the truth of the proposition seems quite certain. This is as far as Euler went in the first paper; the next step, of actual proof, came a year later in the second paper.

Some of the other propositions and their corollaries in the first paper will be reviewed briefly.

I. The number of edges is equal to half of the total number of sides of the faces, since two sides meet at each edge; and, since the number of plane angles of the faces is equal to the number of sides of the faces, the number of edges is also equal to half the number of plane angles. The total number of sides, and hence also the total number of plane angles, must be even, since the number of edges is not fractional. It follows that the number of faces with an odd number of sides, and an odd number of plane angles, must be even.

II, III. Propositions II and III develop relations which are now written

$$2E \geqq 3F \text{ and } 2E \geqq 3V. \tag{1}$$

As to the first, if every face of a polyhedron is triangular the total number of the sides of all the faces would be $3F$ and the number of edges would be half this number; if some faces have more than three sides the total number of sides would be greater than $3F$ and hence the number of edges would be greater than $3F/2$. One can argue analogously for the second relation. (The concepts of edge and vertex having been expressed, it is now not improper to adopt these terms and use the familiar V for number of vertices, F for the number of faces and E for the number of edges in place of Euler's S, H and A, respectively; this will be done from now on.)

IV. Stated in our terminology, Euler's polyhedron formula is

$$V + F = E + 2.$$

V, VI. The above relations are independent of the theorem of Proposition IV. Propositions V and VI combine the relations (1) with it to produce two pairs of inequalities now written as

$$2V \geqq F + 4, \quad 2F \geqq V + 4 \tag{2}$$

and

$$3V \geqq E + 6, \quad 3F \geqq E + 6. \tag{3}$$

VII. A solid body with each face having six or more sides, or with each solid angle formed from six or more plane angles, is not possible. For suppose every face has six or more sides: then the total number of sides would be equal to or greater than $6F$, and hence $2E \geqq 6F$, or $E \geqq 3F$; taken with the second of the inequalities (3), this would result in the absurdity $3F - 6 \geqq E \geqq 3F$. One can argue analogously with respect to the solid angles.

VIII. Propositions VIII and IX are concerned with two different formulas for the sum of the plane angles of all the polygonal faces of a polyhedron, expressed verbally and in equation form. The first is in terms of edges and faces and is written here as (Euler wrote out the words "right angles" which are here expressed as $\triangle$)

$$R = 4(E - F)\triangle.$$

This was proved directly in a simple manner, which was even further simplified by Legendre. Legendre's proof as amplified by using formulas is as follows. The sum of the plane angles of a face with n_i sides is $2(n_i - 2)\triangle$. There are F faces and F such sums, and adding these sums together results in the sum of all the plane angles being equal to $2(\Sigma n_i - 2F)\triangle$. The term Σn_i is the sum of the number of sides of all the polygonal faces, which is equal to $2E$. Hence $R = 4(E - F)\triangle$.

IX. Proposition IX is the main angle sum formula, in terms of the number of solid angles the same as Descartes' formula, and is written here as

$$R = 4(V - 2)\triangle.$$

This is not proven in the first paper; Euler derives it by substitution of the previous angle sum formula (VIII) into the polyhedron formula (IV).

It is not clear how Euler discovered this formula. It could have been discovered by induction (Pólya uses this as another example of discovery by induction but does not assert that Euler did so) and the easily proven angle sum formula (VIII) then derived to use with the polyhedron formula (IV). There was a failure on Euler's part to derive a simple proof of this angle sum formula. It could have been done by following analogies with plane figures, as has been shown (p. 46), but Euler began by showing that such analogy was useless for his objective, and he evidently stayed away from recourse to analogy. A year later he did produce independent proofs of the angle sum formula and of the polyhedron formula, but these were not of a simple nature.

Euler realized the importance of the angle sum formula in his first paper, for if an independent proof of it were found, then his polyhedron theorem could be proved. In the letter to Goldbach which preceded the reading of the paper by a few weeks and which gave a summary of his results, he stated that he found it surprising that these general results in stereometry had not been noticed before by anyone, as far as he knew; then he added that the two important results, referring to the polyhedron formula (IV) and the angle sum formula (IX), were so difficult that he had not yet been able to obtain a satisfactory proof.

The last part of the paper is devoted to his initial objective, which he referred to as Problem 1: to enumerate the genera to which every polyhedron could be referred and to give them names. The pairs of inequalities written above (p. 68), together with the basic theorem (IV), enable the calculation of all triplets of values for V, F and E that a polyhedron could possibly have. To see this readily, pairs (1) and (3) can be combined into one expression

$$2E/3 \geqq (V,F) \leqq (E + 6)/3.$$

For a given $E (\leqq 6)$, integers on or between the outer limits supply all possible values for V and F, which are paired when checked against the main formula. (Euler notes, incidentally, that a polyhedron with seven edges is not possible.) A genus, in Euler's classification, comprises all polyhedra which have the same number of vertices, edges and faces. A table of the genera is presented, arranged first by the number of vertices and then by number of faces and edges and going up to ten vertices. The classification is topological, as is his description of the single species in each of the first four genera. He states that the following genera generally include many species but that the other relevant properties of solids have not yet been developed sufficiently to permit their enumeration.

The second paper, which provided the proofs lacking in the first, was read about a year later. This lapse of time may indicate the difficulty Euler experienced in finding proofs. Each of the two formulas is proven separately and independently of the other, although the same procedure is involved in each. The proofs are quite inelegant and will not be described here. While the proof of the polyhedron formula was accepted by later geometers, for example by Legendre in his *Géométrie,* it has been criticized by Lebesgue [72] as lacking in rigor. The proof has also been termed invalid because of the breadth of the statement "All solid bodies with plane faces...," since there are solid bodies with plane faces to which the theorem does not apply. This criticism is not justified, since the context of the papers shows plainly that he was dealing with convex polyhedra; the etymology and the usage of terms in Euler's day would also indicate convex polyhedra.

Legendre gave a simple proof of Euler's polyhedron theorem in his *Géométrie* of 1794, and developed various consequences of it in an appendix. The proof, specifically limited to convex polyhedra, is based on projecting the vertices and edges onto the surface of a unit sphere with its center in the interior of the polyhedron, from the center of the sphere. The surface is divided into spherical polygons corresponding respectively to the faces; for each face there is a spherical polygon with the same number of sides and angles and they are arranged in the same order. Legendre developed the known formula for the area of a spherical polygon by dividing it into triangles and adding together the areas (cf. Section 5, p. 39). The result is $a = s - 2n\triangle + 4\triangle$, where a is the area, s the sum of the angles, and n the number of sides. There are F such formulas, one for each polygon on the sphere, and these are added together. The sum of the terms on the left is the total area of the sphere, which is $8\triangle$. The sum of the first terms on the right is the sum of all of the angles of the polygons; these angles surround V points on the sphere without gaps or overlapping and, as the sum of the angles about a point on a sphere is $4\triangle$, this term is equal to $4V\triangle$. The sum of the n's is the sum of all the sides of all the polygons, which is equal to the sum of the sides of the faces of the polyhedron and hence to $2E$. The result of the addition is

$$8\triangle = 4V\triangle - 4E\triangle + 4F\triangle,$$

which gives $V + F = E + 2$.

Legendre's text became very popular and went through many editions, the eighth appearing in 1809 and the twelfth in 1823, and it was translated or adapted into other languages. It served to make Euler's theorem and the simple proof widely known among mathematicians, so that it was included in other texts which covered solid geometry, and would be further considered by geometers.

Of the many geometers who discussed Euler's theorem we will briefly note the contributions of some of the earlier ones. In 1810 Poinsot[73] called attention to the fact that Legendre's proof went beyond convex polyhedra, as it applied without change to polyhedra with re-entrant solid angles, provided it were possible to find an interior point to be the center of a sphere onto which the faces could be projected without overlapping. In the years 1811 and 1813 Cauchy, Lhuilier and Gergonne[74] came up with about a half dozen new and different proofs, starting a flood of proofs. Lhuilier also called attention to a variety of non-convex solid bodies to which Euler's formula did not apply and developed variations of the formula for them. Steiner gave another proof (1826); he also revived Euler's problem of enumeration (1828),[75] which thereafter received the attention of a number of mathematicians. And there were many others, notable among which was that of von Staudt (1847).[76] He

stated a purely topological necessary and sufficient condition for a polyhedron to satisfy Euler's theorem and gave a simple and elegant proof of the
theorem, also purely topological. Among the early topologists who gave
proofs and extensions may be mentioned Listing (1862), Möbius (1863), and
Jordan (1866).[77] We have already quoted (p. 63) Poinsot's remark of 1858
about the difficulty of the theory of polyhedra. Lebesgue said, in an
unpublished notebook, that, "The first important notions in topology were
acquired in the course of the study of polyhedra."[78]

9 Descartes and Euler

Our analysis of and comments on the manuscript in Sections 6 and 7 did not mention Euler's work, and the summary of Euler's work in Section 8 did not mention Descartes; this was done in order that all comparative remarks might be considered at the same time.

There is a good deal of parallelism between Descartes' notes and Euler's work. Each is an original attempt at a general theory of polyhedra, and each departed from the prior treatment of this subject in a significant manner; the comments on the "state of the art" in 1630 made in Section 7 (p. 61) apply with equal force to 1750. As to subject matter, Descartes' Proposition 2 is also given by Euler (his Proposition IX) and the two inequalities of Proposition 3 are also stated by Euler (in his Proposition V), who has other pairs of inequalities as well. Descartes' Proposition 5, giving the sum of the plane angles in terms of the number of faces and plane angles, finds its counterpart in Euler's Proposition VIII, which gives this sum in terms of the number of faces and edges. And Descartes' Proposition 6, a relation between the number of plane angles, the number of faces and the number of solid angles, finds its counterpart in Euler's Proposition IV, the relation between the number of edges, faces and vertices, the famous Euler formula or theorem. The last two mentioned propositions of Euler can be derived from the corresponding propositions of Descartes by introducing the notion of edges of a polyhedron (which Descartes does not have), and eliminating plane angles from consideration (which Descartes does not intimate); this matter is considered below. Euler does not have Descartes' Proposition 4 dealing with regular solids, nor Descartes' Proposition 1, on the sum of the exterior solid angles, but does have a number of propositions not in the Descartes' notes; the absence of any reference to exterior solid angles in Euler may be due to the different approaches.

The approaches in the two works are quite different. Descartes' is based upon analogy with plane figures and shows little or no signs of induction, whereas Euler's is based upon induction, with analogy cast aside (although an analogy is mentioned at one point). The central element in Descartes is the plane angle: Propositions 2, 5 and 6 relate to plane angles, and 6 is only used to determine the deficiency in the number of plane angles from the maximum number possible; Propositions 3 and 4 are derivable from considerations of plane angles; and various statements and examples involve plane angles. On the other hand, Euler introduced a new element, the edge, which became central in his work, even though he gives two angle sum formulas.

Euler's two papers contain a good deal more than the Descartes notes. This is natural, since the Euler papers form a finished memoir whereas the notes were only the first steps of a discontinued project. But Euler also had the advantage of working 120 years later, when the level of general mathematical knowledge was much higher than in 1630 (owing to the work of Descartes and others) and furthermore he had a notion of topology (first expressed by him 15 years earlier), which notion did not exist in 1630.

With respect to the relation of the Descartes' manuscript to Euler's formula, two statements can be made:

1. The manuscript does not disclose Euler's formula directly. By Euler's formula is meant his relation that the sum of the number of faces and the number of vertices is equal to the number of edges plus two ($F + V = E + 2$).

2. The manuscript does disclose several statements from which Euler's formula can be derived (Descartes' Proposition 6 can even be called an analogue of Euler's formula); this follows from the manuscript with the addition of present-day knowledge of Euler's formula.

The simplest ways of deriving Euler's theorem from the manuscript are mentioned here (some others will be noted later). First, the union of the two meeting sides of adjacent faces of the polyhedron is taken as the element of the polyhedron to be considered and given a name, "edge," and a symbol, E. Then the statement of Paragraph 12 of the manuscript can be expressed as: the number of plane angles is equal to twice the number of edges, or $P = 2E$. $P = 2E$ can be substituted into Proposition 5, forming $\frac{\Sigma}{\Delta} = 4E - 4F$ (which is Euler's Proposition VIII) and this combined with Proposition 2($\frac{\Sigma}{\Delta} = 4S - 8$) gives Euler's formula (after dividing by 4). Alternatively, $P = 2E$ can simply be substituted into Proposition 6, giving $2E = 2F + 2S - 4$, which results in Euler's formula after dividing by 2.

The above is so simple that some authors have asserted as a fact that Descartes knew or was aware of Euler's theorem. Since we do not actually know what was in his mind other than what is stated in the manuscript, such a conclusion could only properly be stated in terms of probability rather than fact. The question, "Did Descartes know Euler's theorem?" or, alternatively stated, "Was Descartes aware of Euler's theorem?" can be answered only: "Yes, probably" or "No, probably," with or without further qualification.

The least extravagant type of reasoning leading to the unqualified factual affirmation made by some authors can be expressed by the following propositions:

1. X knew A.
2. B is a logical (simple, direct, immediate) consequence of A.
3. Therefore, X knew B.

But the third term does not necessarily follow from the first two; other considerations must come into play before even a plausible inference could be drawn.[79]

But even if a plausible inference could properly be made, it does not necessarily follow that the inference was in fact made. Hadamard, in his discussion of this matter,[80] refers to "the failure of a research scholar to perceive an important immediate consequence of his own conclusions" as a "paradoxical" failure. This psychological failure he attributes to concentration or too narrowly a focusing on the problem or subject at hand. He recites four personal instances, the second of which we quote:

> My next work was my thesis. Two theorems, important to the subject, were such obvious and immediate consequences of the ideas contained therein that, years later, other authors imputed them to me, and I was obliged to confess that, evident as they were, I had not perceived them.

Hadamard gives instances of similar experiences by others, and concludes with the remark, followed by an instance involving Pascal, "It is probable that many scholars, if not all of them, can remember similar experiences. It is a comforting thing to think that the same may happen to the greatest ones."

But Hadamard was referring to contemporaneous or substantially contemporaneous events and knowledge. With respect to the present question we have the situation (referring to the above schema) that X knew A in 1630 and B became known in 1750, and the fact that B could be derived from A became known in 1860.

Quite a number of books and articles contain such statements as that Descartes anticipated Euler's theorem, that the Descartes manuscript discloses Euler's theorem, that Descartes knew, or was aware of, Euler's theorem, etc., referring to Euler's theorem relating the edges, faces, and vertices of polyhedra, without more. I have noticed such statements in half a dozen English language general histories of mathematics and in a like number of mathematical books which make historical remarks relating to their subject. There are undoubtedly many others, but no effort has been made to seek them out. Instead, a dozen of the pronouncements on this matter are gathered in the following *addendum*. These are not intended as "authorities" (since the only authority is the manuscript itself). They range from the respectable to the ridiculous. The list, which is not intended to be complete, is arranged in chronological order.

Addendum to Section 9

1. *Prouhet, 1860.* Prouhet (I) was the first to comment on the relation of the Descartes manuscript to Euler's theorem. One of the reasons he gave for the importance of Descartes' theorem (Proposition 1) was that Euler's theorem could be derived from it. After showing how Descartes' theorem could be demonstrated and how the angle sum formula (Proposition 2) could be derived from it, he showed how Euler's formula could be found. His method consisted in deriving Euler's Proposition VIII (in substantially the same way that Euler did) and combining it with Proposition 2. He concluded the first note with the statement that Descartes did not express Euler's theorem but "the exact rules he gave on the number of elements of certain solids, show that he had carried the consequences of the equality [Proposition 1] very far." He considered it obvious that Descartes was dealing only with convex polygons and convex polyhedra.

In his full treatment which came later in the year (Prouhet II), Prouhet again emphasized the importance of the manuscript because of "a very beautiful and general theorem which ought to be placed henceforth at the head of the theory of polyhedra and of which Euler's theorem, regarded up to the present as fundamental, is no more than a simple corollary." This referred again to Proposition 1, as stated in the next column of the paper.

2. *Baltzer, 1861, 1862.* The first Baltzer reference[81] is a short note on two topics, the first of which is the Descartes manuscript. He refers to Foucher de Careil and Prouhet I only. He first quotes the paragraph of the manuscript on the exterior angle sum (Proposition 1) and then the plane angle sum theorem (Proposition 2). The latter is called the fundamental theorem (Grundgesetz) of polyhedrometry and corresponds with Euler's second theorem (Euler's Proposition IX) which the latter discovered more than 100 years later and which Euler regarded as the basis for another. He quotes the sentence from Euler in which Euler stated that it (his Proposition IX) was related to his main theorem (his Proposition IV), since if one were demonstrated the other would be demonstrated. Baltzer sets up the angle sum formula, Descartes' Proposition 2 and Euler's Proposition IX, as the fundamental theorem of polyhedra. In his concluding paragraph Baltzer states:

> Therefore it is undoubted that the discovery of the fundamental theorem (Grundgesetz) of polyhedrometry also belongs to the brilliant achievements which glorify the name of Descartes, and that Euler must henceforth share the honor of that discovery with his great predecessor (the founder) of modern analysis.

When quoted out of context, or without stating what theorem is meant, this obviously gives a false impression of what Baltzer said.

Baltzer also quotes the paragraph with the second angle sum theorem (Paragraph 11, Proposition 5) and derives Euler's theorem. This is done by combining the two angle sum formulas and then replacing the number of plane angles by the corresponding number of edges; the result, he states, corresponds with the first Euler theorem on the number of faces, vertices and edges. He does not assert that this derivation is in the manuscript or that Descartes was aware of it.

Baltzer's book of 1862[82] gives Euler's theorem (the formula connecting the number of vertices, edges and faces) in the text, with the following footnote:

> This basic theorem of polyhedrometry (which perhaps was known in antiquity, as Archimedes was able to enumerate the complete series of semiregular polyhedra), first appears in a fragment of Descartes published in 1860, *Oeuvres inédites de Descartes* by Foucher de Careil, II, p. 214. Compare the note of the present author in *Monatsbericht der Berl. Acad.* 1861, p. 1043. The theorem was first made known by its rediscoverer, Euler, in 1752, *Nov. Comm. Petrop.* 4 p. 109, and proven p. 156.

3. *Becker, 1869.* Becker[83] is concerned with extensions of Euler's formula beyond convex polyhedra and also develops formulas involving plane angle sums. He notes Descartes' theorem on the angle sum (Proposition 2), and misunderstanding Prouhet, states, "Mr. Prouhet concluded from it that Descartes must have already known Euler's theorem, but stopped short of proof of that theorem." Becker does not agree, stating, "This conclusion seems to me in fact patriotic, but not very well founded (gerechtfertigt)." Several reasons are given; he points out that neither theorem is a logical consequence of the other, as each is independent of the other (that is, can be derived independently).

4. *De Jonquières, 1890.* The principal contender that Descartes was aware of Euler's theorem and expressed it is de Jonquières. In his first note[84] he states that two theorems are found in the notes, from which the remarkable relation between the number of faces, vertices and edges of a polyhedron flows intuitively. These are the two angle sum formulas (Propositions 2 and 5), which he gives. By substituting twice the number of edges for the number of plane angles in the second formula, and combining the two, Euler's relation is produced. "It cannot be denied then that he knew it, since it is a deduction so direct and so simple, we say so intuitive, from the two theorems that he had just stated." He concludes with the peroration, "One must then, without lessening the merit of Euler, who independently found it later, add this new jewel to the crown of our great compatriot."

In the second note[85] de Jonquières asserts in the opening paragraph that Descartes not only knew and applied Euler's formula, (as he had previously remarked), but had also stated it explicitly. He refers to the formula for the number of plane angles in terms of the number of faces and solid angles (Proposition 6). By substituting twice the number of edges for the number of plane angles, Euler's formula is obtained "explicitly expressed." The logic can be stated as: (1) X expressed A; (2) B is an immediate consequence of A; (3) therefore X expressed B.

De Jonquières' first note states generally that Descartes applied Euler's theorem to numerical examples. His second note gives two so-called applications of Euler's theorem. The first is the problem posed in Paragraph 13 of the manuscript (see Section 6, p. 55). The derivation of the number of solid angles in the problem as 6 is a "necessary consequence" of Euler's theorem, since, given that $F = 5$ and that the number of plane angles is 18 (from which the number of edges is 9), it follows by substitution in Euler's formula that $S = 6$. The second application is the pair of formulas used for determining the number of regular polyhedra (see Section 6, Paragraph 7, p. 50), which can be derived from Euler's formula by using the definition of the regular bodies and the notion of edges. These applications are given as confirmations of de Jonquières' thesis. The note ends with the statement that since Descartes was prior in fact, though not in publication, it is equitable henceforth to associate his name with that of Euler and to call the formula the Euler–Descartes relation.

The third note[86] is a description of the memoir presented to the Académie the same day. The memoir repeats or restates the various assertions previously made, with some amplifications, which we need not consider here. De Jonquières quotes the last paragraph of Baltzer's note, of which he had not been previously aware, as supporting his conclusion.

5. *Killing and Hovestadt, 1913*. This book on the teaching of mathematics has a section on Euler's theorem, the opening historical paragraphs of which are quoted in full:[87]

> The noteworthy theorem which now bears Euler's name was, indeed, not clearly expressed by the Greek mathematicians, but according to Baltzer's conjecture, supported by important grounds, did not remain unknown. He finds its clear expression in the posthumous paper of Descartes, which was first printed in the year 1860. Without having knowledge of it, Euler discovered the theorem anew and provided a proof. Here, in his discovery published in the writings of the Petersburg Academy, he brought the theorem to general knowledge and it was soon accepted in textbooks.
>
> The theorem consists of several relations, of which two are particularly important. In order to express them in formulas, we designate the number of vertices by e, the edges by k, and the faces of a polyhedron by f, and designate the smallest number of triangles into which the surface can be divided equal to D. Then the theorem is affirmed, that is,
>
> (1) $e + f = k + 2$
> (2) $D = 2(e - 2)$.
>
> The first equation becomes associated with Euler's name in the narrowest sense, when one speaks of Euler's theorem of solid geometry, so that one thinks only of equation (1). On the other hand, Descartes places equation (2) in the first place and derives equation (1) from it.

It is needless to comment on this performance.

6. *Lebesgue, 1924*. Lebesgue's paper[88] is on the first two demonstrations (by Euler and Legendre) of Euler's theorem. He regards Legendre's demonstra-

tion as the first rigorous proof. Euler's and Legendre's proofs, as well as the Descartes' manuscript, are discussed with suggestions as to how they arrived at their respective results. Descartes' approach was metric while Euler's was topological. Lebesgue's paper takes an additional view, based on Pólya's analysis of the role of induction and analogy in the discovery of mathematical theorems.

As to Descartes' knowledge of Euler's theorem, Lebesgue states emphatically, "...I am not at all in agreement with those who claim that one can attribute to Descartes the theorem of Euler. Descartes did not enunciate the theorem; he did not see it" (p. 319). Again he indicates (p. 320), "That Descartes came so close to the theorem without seeing it" could be due to the fact that he was young at the time, "but Leibniz, who found Descartes' notebook sufficiently interesting to copy it...did not perceive, in Descartes' notebook, the theorem of Euler so fundamental in analysis situs."

In one of Lebesgue's unpublished notebooks (quoted by Pont[89]) appears the statement: "The first important notions in topology were acquired in the course of the study of polyhedra...In 1890, de Jonquières, in a series of notes in C.R., created the legend that this theorem was due to Descartes. Descartes did not express Euler's theorem; as for topology, he by no means surmised it;..."

7. *Cajori, 1929.* Cajori first states, of Euler's relation between the edges, vertices and faces of a convex polyhedron, that, "The theorem was first enunciated by Descartes, but his treatment of it was not published until 1859 and 1860."[90] Then, in the same paragraph, he gives Descartes' relation between the number of plane (polygonal) angles, faces and solid angles and states, "As the sum of the polygonal angles is twice as great as the number of edges, it is evident that Descartes' phrasing is equivalent to Euler's statement."

8. *Steinitz and Rademacher, 1934.* The authors of this book on polyhedra state[91]:

> Euler's desire [to find a proof of the angle sum formula, his Proposition IX and Descartes' Proposition 2] was actually fulfilled by Descartes about a hundred years before he [Euler] formulated it. In fact he [Descartes] found and proved formula (9) [the angle sum formula] and, starting from it, also Euler's theorem (1) [Euler's theorem].

9. *Fréchet and Fan, 1946.* Pont[89] criticizes two French books[92] for falsely asserting that Hilbert and Cohn-Vossen have shown that Euler's Theorem (connecting the number of vertices, faces and edges of a convex polyhedron) was due to Descartes. In fact the later author, A. Delachet, derived his misinformation from the earlier book, *Initiation to Combinatorial Topology* by Maurice Fréchet and Ky Fan.[93] This beautiful gem of exposition is marred by the treatment of Euler's theorem. Section 11 is headed "Descartes' theorem," and the opening paragraph states:

"Well known is the famous *Descartes' formula* (mentioned by Poincaré in the quotation on page vi), often attributed to Euler[1]:

$$(1) \quad n_v - n_e + n_f = 2. \text{"}$$

The footnote attached to this sentence reads: "[1]On the matter of priority to Descartes, see: W. Killing and H. Hovestadt [16], page 268; D. Hilbert and S. Cohn-Vossen [11], page 254."

Page 254 is the beginning of section 44, "Polyhedra", of Hilbert and Cohn-Vossen.[94] After describing polyhedra, the authors state:

> Die anzahlen der Ecken, Kanten und Flachen eines simplen Polyeders stehen zueinander immer in einer wichtigen Beziehung, die nach ihren Entdecker der EULERsche Polyedersatz genannt wird.

In the English edition:

> There is an important relation between the number of vertices, edges, and faces of a simple polyhedron. It is called *Euler's Formula for Polyhedra*, after its discoverer.

Thereafter the work consistently refers only to Euler; Descartes is never mentioned. Incidentally, the quotation from Poincaré referred to does not mention Descartes. The Killing and Hovestadt reference has been cited above (p. 77).

The false reference to Hilbert and Cohn-Vossen in Fréchet and Fan is undoubtedly due to a slip on the part of the authors. Thence it was uncritically accepted by Delachet, and unfortunately also by the English translator of Fréchet and Fan (whose copious notes are otherwise up to the high standard of the original).

10. *Pólya, 1954, 1965.* Perhaps the person who has studied the papers of Euler and the notes of Descartes more intensively than anyone else is George Pólya. He frequently refers to them, particularly Euler, in two books[95] and various papers dealing with the role of guessing, analogy, and induction in mathematical discovery, in the solving of problems, and in teaching. He uses various propositions of Euler, and also Descartes, as illustrative examples, and sets the proof of some as problems; a few of these have already been referred to (pp. 37, 45). As to the Descartes notes he states, "These notes treat of subjects closely related to Euler's theorem: although the notes do not state the theorem explicitly, they contain results from which it immediately follows."[96] He is definitely of the opinion that the notes do not state Euler's theorem (in a letter in response to an inquiry he stated that he had read the Descartes manuscript many times and has just read it again and still does not find it). Pólya does not discuss whether Descartes was aware of Euler's theorem (he does not indulge in historiographic vices), but he evidently would not think so for he pointed out that it was Euler who invented the concept of the edges of a polyhedron and gave them a name.

11. *Pont, 1973.* Pont devotes an early section of his history of algebraic topology[97] to the Descartes manuscript. Referring to the manuscript, he says: "Its principal interest resides in the fact that it contains a proposition from

which Euler's theorem on polyhedra is an almost immediate consequence. Two radically opposed interpretations are accordingly possible: one attributes Euler's theorem to Descartes, the other affirms that he did not see it. The object of the present section is the study of these two appreciations." After consideration of various formulas in the manuscript and their derivations, and the arguments of de Jonquières, he comes to the same conclusion as Lebesgue.

12. *Lakatos, 1976*. This book on the logic of mathematical discovery utilizes the theme of proofs of Euler's theorem throughout. The discussion begins with a "proof" based on Cauchy's proof of 1813, but so modified that defects can be found. The first mention of the formula $V - E + F = 2$ carries a long footnote, the first paragraph of which reads: "First noticed by Euler [1758a]. His original problem was the classification of polyhedra, the difficulty of which was pointed out in the editorial summary: 'While in plane geometry polygons *(figurae rectilineae)* could be classified very easily according to the number of their sides, which of course is always equal to the number of their angles, in stereometry the classification of polyhedra *(corpora hedris planis inclusa)* represents a much more difficult problem, since the number of faces alone is insufficient for the purpose." The second paragraph begins with the sentence quoted above p. 66, and then refers to the Descartes manuscript and the view, stated as recently generally accepted, that Descartes anticipated Euler. The footnote ends:

> It is true that Descartes states that the number of plane angles equals $2\phi + 2\alpha - 4$ where by ϕ he means the number of faces and by α the number of solid angles. It is also true that he states that there are twice as many plane angles as edges *(latera)*. The conjunction of these two statements of course yields the Euler formula. But Descartes did not see the point of doing so, since he still thought in terms of angles (plane and solid) and faces, and did not make a conscious revolutionary change to the concepts of 0-dimensional vertices, 1-dimensional edges and 2-dimensional faces as a necessary and sufficient basis for the full topological characterisation of polyhedra.

The translating of "latera" as "edges" instead of "sides" is anachronistic; Lakatos himself states that Euler invented the concept of edge. But even with this addition to the manuscript he does not believe that Descartes joined the two statements to produce Euler's formula.

Part Three
Number Theory: Polyhedral Numbers

10 The Figurate Numbers of the Greeks

The figurate numbers of the Greeks go back to the time of Pythagoras and are frequently referred to by Greek authors. The natural way to represent numbers was by a set of units, shown by dots in sand or by pebbles, and these were arranged in patterns of geometrical figures. In writing, a dot or the letter α would be used for each unit. The figurate numbers and various relations and problems concerning them form a substantial part of Greek number theory (arithmetic). Surviving works which treat figurate numbers are by Nicomachus of Gerasa (c. 100 A.D.), Theon of Smyrna (c. 130 A.D.),[99] Diophantus of Alexandria (c. 250 A.D.),[100] and Iamblichus (c. 283–330 A.D.).[101] Summaries are given by Heath and Dickson, and some relevant extracts by Cohen and Drabkin.[102] This section will review some of this material, with some added matter, before presenting the text of the second part of the manuscript, which deals with figurate numbers.

Nicomachus divides numbers into linear or one-dimensional, plane or two-dimensional, and solid or three-dimensional. Plane numbers are represented by two-dimensional figures and are divided into two classes; (1) polygonal numbers, in which the units (dots) are arranged in the form of equilateral polygons (see below), and (2) oblong numbers, the product of two factors (presumably not including one as a factor), which would be shown as rectangular arrays of units according to the two factors as sides. Solid numbers are likewise divided into two classes; (1) pyramidal numbers (see below p. 89)—he also considered truncated pyramids—and (2) numbers which are the product of three factors (if the three factors are equal the numbers are cubic, if two are equal and the third smaller they are bricks, and if the third is larger they are beams; if all three factors are different the numbers are scalene, with unequal sides). All the above are treated in chapters 6 to 17 of Book II of Nicomachus; numerous special terms are used for particular types.

(1) *Polygonal numbers.* Figures 9–12 show the beginnings of the series of triangular, square, pentagonal and hexagonal numbers, respectively. Note that for every polygon the first number in its series is simply one. The first line

under each figure gives the number of dots along a side of the polygon; this number will be referred to in the general case as *n,* and is the radix or base of various formulas. It is represented in the Descartes manuscript by the cossic symbol for the radix, $\mathcal{R}$.

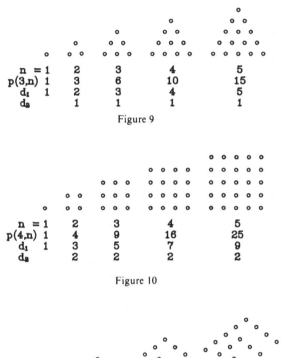

n = 1	2	3	4	5
p(3,n) 1	3	6	10	15
d₁ 1	2	3	4	5
d₂	1	1	1	1

Figure 9

n = 1	2	3	4	5
p(4,n) 1	4	9	16	25
d₁ 1	3	5	7	9
d₂	2	2	2	2

Figure 10

n = 1	2	3	4	5
p(5,n) 1	5	12	22	35
d₁ 1	4	7	10	13
d₂	3	3	3	3

Figure 11

The second line under each figure gives the numerical value of the corresponding polygonal number. These numbers will be represented here by the notation $p(a,n)$, where a is the number of sides of the polygon and n is the number of dots along a side, so $p(3,n)$, $p(4,n)$, $p(5,n)$ and $p(6,n)$ are written at the beginning of each respective second line.[103] The third and fourth lines will be referred to later.

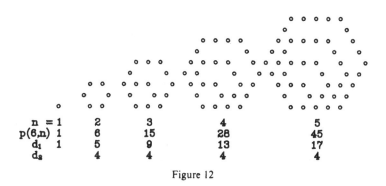

n = 1	2	3	4	5
p(6,n) 1	6	15	28	45
d₁ 1	5	9	13	17
d₂	4	4	4	4

Figure 12

For certain purposes, which follow later, the triangular numbers are represented by an elongated isosceles triangle as in Fig. 13(a) or by a form elongated and sheared, Fig. 13(b). Figure 13(c), showing another form, illustrates the formation of a square number from two triangular numbers; in general, $p(4,n) = p(3,n) + p(3,n - 1)$.

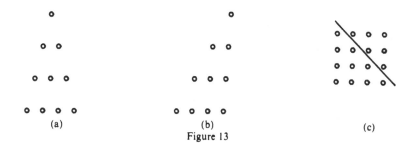

(a)   (b)   (c)

Figure 13

Hypsicles (c. 150 B.C.) as quoted by Diophantus,[104] gave the following definition of polygonal numbers:

> If there are as many numbers as we please beginning with one and increasing by the same common difference, then when the common difference is 1, the sum of all the terms is a triangular number; when 2, a square; when 3, a pentagonal number. And the number of angles is called after the number exceeding the common difference by 2, and the side after the number of terms including 1.

Thus if you start with a simple arithmetical series (which we call a "series of the first order," in which the first differences of the terms are constant) beginning with 1, the sum of the first n terms is a polygonal number with the number of dots on a side equal to n. If the number of sides (angles) of the polygon is a, the common difference is $a - 2$. These relations are illustrated in Figs. 9–12. In each case the third line under the figure is the arithmetical series and the fourth line the common differences. If, instead, we start with the polygonal numbers (line 2), the first differences d_1 are given in line 3 (a zero is imagined as preceding the 1 in line 2), and the fourth line gives the common second difference. Each series of polygonal numbers is an arithmetical "series of the second order" (in which the second differences are common).

Since each series of polygonal numbers is formed by the successive summation of the terms of a simple arithmetical series, the general term for each series can be readily found from the elementary properties of arithmetical series. Given the first term and the common difference, the nth term is $(n-1)$ times the common difference added to the first term, and the sum of the first n terms is n times the sum of the first and last terms divided by two. The general terms for the four series illustrated in Figs. 9–12 and the three following are

$$
\begin{aligned}
p(3,n) &= (n^2 + n)/2, \\
p(4,n) &= n^2, \\
p(5,n) &= (3n^2 - n)/2, \\
p(6,n) &= 2n^2 - n, \\
p(8,n) &= 3n^2 - 2n, \\
p(10,n) &= 4n^2 - 3n, \\
p(12,n) &= 5n^2 - 4n.
\end{aligned}
$$

These formulas are used by Descartes in arriving at results stated in the manuscript. Only some are written down by him; the n, of course, is written as the cossic symbol for the radix or base, ℛ, and the n^2 as the cossic symbol for zenzus, the radix squared. He called the formulas the "exponents" of the corresponding general polygonal number. His method of deriving them is given later (p. 94).

Formulas of the kind given above for the general term of individual series of polygonal numbers were known before Descartes' work. Faulhaber gives cossic equations for the series of triangular, square, pentagonal and hexagonal numbers (see Section 12, p. 117). A completely general formula for a polygonal number corresponding to a polygon with any number of sides and any base would not have been written at the time, rather, a verbal description would have been used.

To obtain a completely general formula, we follow Hypsicles in taking $a - 2$ as the common difference where a is the number of sides of the polygon: then the nth term of the arithmetical series is $1 + (n - 1)(a - 2)$, and the sum of the first n terms (n times the sum of the first and last terms divided by two) reduces to

$$p(a,n) = (a - 2)n^2/2 - (a - 4)n/2. \tag{1}$$

If d is taken as the common difference the formula can be written in the simpler form

$$p((d + 2),n) = n + n(n - 1)d/2 = dn^2/2 - (d + 2)n/2.$$

Diophantus (tr. Heath, pp. 249–254) gave essentially the same two formulas in verbal form, with complicated geometrical proofs. He also gave an expression for n in terms of the other quantities. The fragment "On Polygonal Numbers" breaks off after the first steps of a solution of the problem: "Given a number, to find in how many ways it can be polygonal."

Nicomachus presented a table of the polygonal numbers up to the heptagons and extending to the tenth member of each series. This table is repeated here with added headings and side notations. He called attention to a particular relation, by examples and in general terms. Note that the ninth hexagonal number, 153, is equal to the ninth pentagonal number immediately above it, 117, plus the eighth triangular number, 36. This is true in general: the nth a-gonal number ($a > 3$) is equal to the nth $(a - 1)$-gonal number plus the $(n - 1)$th triangular number. (For the first column, $n = 1$, a zero is assumed to precede the first triangular number.) This relation is expressed in the notation used here as the recursive formula $p(a,n) = p(a - 1,n) + p(3,n - 1)$. Each *column* of the Nicomachus table is an arithmetical series of the first order, and the common difference is the first number in the preceding column.

$n =$	1	2	3	4	5	6	7	8	9	10
$p(3,n)$ Triangles	1	3	6	10	15	21	28	36	45	55
$p(4,n)$ Squares	1	4	9	16	25	36	49	64	64	81
$p(5,n)$ Pentagonals	1	5	12	22	35	51	70	92	117	145
$p(6,n)$ Hexagonals	1	6	15	28	45	66	91	120	153	190
$p(7,n)$ Heptagonals	1	7	18	34	55	81	112	148	189	235

From this the general term is readily seen to be

$$p(a,n) = p(3,n) + (a - 3)p(3,n - 1),$$

which reduces to the formula (1). Nicomachus considered the triangular numbers basic for polygonal numbers, since any one can be formed by adding triangular numbers, at least arithmetically if not geometrically as shown in Fig. 13(c).[105]

Numerous relations among the polygonal numbers and problems concerning them were studied by the Greeks and later writers. Dickson in Chapter 1 of Volume II of his *History of the Theory of Numbers* lists 200 names. But only a few of these are of any concern here. Two related geometrical aspects need to be considered, gnomons and division into triangles.

(2) *Gnomons and division into triangles.* Figures 9–12 illustrate the fact that each polygonal number after the first two is a nest of linear polygons of the same number of sides having one angle in common. Thus each polygonal number can be obtained from the preceding one of the same number of sides by adding a partial new border. These additions are called gnomons and are illustrated in Figs. 14–17. Consider first Fig. 15, the square numbers: each square is formed from the preceding one by adding a row of n dots along two adjacent sides. In these additions the dots have been connected by lines, showing their shape, which is somewhat similar to a carpenter's square (particularly when the square number is shown as a square array of unit

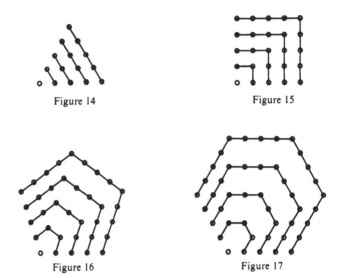

Figure 14   Figure 15

Figure 16   Figure 17

squares with a dot in the center of each and two outer sides shaded, as is sometimes done). The carpenter's square was called a "gnomon" in Greek and is believed to be the origin of the name for these figures. The gnomons in the other three figures are also connected by lines; in the case of the triangles they are simple lines, for the pentagons they have 3 sides and 2 corners, and for the hexagons 4 sides and 3 corners. The successive gnomons are numbered in the figures; number 1, the single dot, is included as a gnomon to complete the series; it is added to zero to form the first polygonal number.

Several features common to all gnomons (excluding the 1) should be noted. In each case the gnomon is a polygon with one angle (including its two sides) removed (Descartes called it a vacant angle); the number of sides is two less than the number of sides of the polygon,[106] and the number of corners is three less than the number of angles of the polygons. If a is the number of sides of the polygon the gnomon has $(a - 2)$ sides and $(a - 3)$ corners. (The word corner is used rather than angle for a reason which will appear in the next section.)

The number of dots in a general gnomon is readily seen to be $(a - 2) n - (a - 3)$. Each of the $(a - 2)$ sides, considered separately, has n dots, therefore $(a-2)n$, but the dots at the corners have thereby been counted twice and hence $(a - 3)$ dots must be subtracted for the $(a - 3)$ corners. This is the recursive method used by Descartes in his tables of polygonal numbers.

The gnomons are obviously the first differences of the terms of a series of polygonal numbers with the same number of sides, the simple arithmetical series from which the polygonal series is derived. This is shown by the line labelled d_1 in each of Figs. 9–12.

Lines drawn from the common corner of the nested set of polygons which form a polygonal number, to each of the other corners, divide the figure into triangles as shown in Fig. 18(a) for the pentagon and Fig. 19(a) for the hexagon. These are related to the gnomons: there are as many triangles as

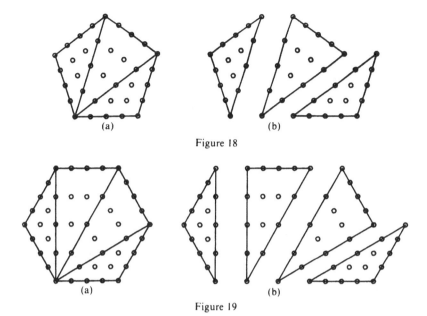

Figure 18

Figure 19

sides in the gnomon, $(a-2)$, and as many sides common to two triangles as the gnomon has corners, $(a-3)$.

The number of dots can be counted from the triangles considered individually. Figures 18(b) and 19(b) show the triangles of the two divisions separated into individual triangles; each of these represents a triangular number with the same radix or base (n) as the polygonal number itself. The total number of dots is

$$p(a,n) \;=\; (a-2)(n^2+n)/2 - (a-3)n.$$

The term $(n^2+n)/2$ is the number of dots in one triangle, which is multiplied by $(a-2)$, the number of triangles; this counts the $(a-3)$ sides twice and hence $(a-3)n$ dots must be subtracted. This formula represents a theorem given by Descartes; it will be recognized as a rearrangement of the terms of the general formula (1).

Figure 20 represents a combination of gnomons and triangles using the pentagon as an example, suggested by Iamblichus' statement. Starting with five points or dots, arranged in the form of a regular pentagon, rays are drawn from one point through each of the other four, rays a, b, c, d in Fig. 20. To form the next pentagonal number a dot is added to each ray, the three dots on each ray being equally spaced on the ray. A gnomon is formed by adding another dot centrally between each adjacent pair of added dots. The next pentagonal number is formed as shown in the figure, and so on. Figure 20 is in effect a combination of Figs. 16 and 18(a).[107]

(3) *Pyramidal numbers.* The Greeks also recognized that if the terms of a series of polygonal numbers were summed, the successive sums formed a series of pyramidal numbers. They were treated by several of the Greek

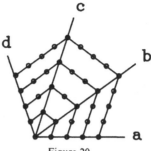

Figure 20

mathematicians who have been mentioned (Nicomachus, Theon, Iamblichus), and others.

Pyramidal numbers are formed by piling up the planes of successive polygonal numbers (see Figs. 9–12), equally spaced and arranged so that the dots form a pyramid (this is essentially the way Nicomachus described them). The lateral faces of the pyramids are congruent triangles with the number of dots along each side equal to the number along each side of the base. Except in the case of the triangular, square or pentagonal base, these lateral triangles cannot be linear equilateral triangles, but are isosceles. For certain purposes, as will appear in the next section, the pyramids can be stretched, or stretched and sheared, while retaining various properties, analogous to the treatment of triangles shown in Fig. 13.[108]

The individual terms of a series of pyramidal numbers can be readily obtained by summing the terms of the corresponding series of polygonal numbers, and the formula for the general term can be derived from the first four. In the following tabulation the third line gives the first four general polygonal numbers and the second line the successive sums of these.

$n =$	1	2	3	4
$P(a,n)$	1	$a + 1$	$4a - 2$	$10a - 10$
$d_1 = p(a,n)$	1	a	$3a - 3$	$6a - 8$
d_2	1	$a - 1$	$2a - 3$	$3a - 5$
d_3		$a - 2$	$a - 2$	$a - 2$

The second and third differences, which are the first and second differences of the polygonal numbers, are also listed. The pyramidal numbers are represented here by $P(a,n)$, where a is the number of sides of the base and n is the radix, the number of dots along each side of the base and the lateral faces.

The pyramidal series is an arithmetical series with constant third differences, which we call a "series of the third order," and the formula for the general term is a cubic polynomial in n. Assuming a cubic with unknown coefficients, four linear equations are formed by substituting the first four values of $P(a,n)$ and the values of the coefficients thus determined.[109] The general formula is

$$P(a,n) = [(a - 2)n^3 + 3n^2 + (5 - a)n]/6. \qquad (2)$$

Formulas for individual series of pyramidal numbers were known before Descartes; Faulhaber gives the "cubicossic" equations of pyramids with bases of 3, 4, 5, 6, 13, 50, 154 sides.

The terms of equation (2) can be rearranged to the following form:

$$P(a,n) = (p(a,n) + n/2) \cdot (n + 1)/3.$$

This equation was given by Descartes, in verbal form, who used it in his calculations, as will be shown in the next section.

Except for the series of pyramidal numbers (which include the tetrahedral numbers) and the parallelopipedal numbers of various names (including the cube), the Greeks do not appear to have had polyhedral numbers nor any general concept of polyhedral numbers.

11 Translation and Commentary, Part II

18 Omnium optime formabuntur solida per gnomones superadditos uno semper angulo vacuo existente, ac deinde totam figuram resolvi posse in triangula. Unde facile agnoscitur omnium polygonalium pondera haberi ex multiplicatione trigonalium per numeros 2./3./4./5./6 etc., et ex producto si tollantur 1./2/3/4 radices, etc.

19 Ut: Tetragonalium pondus fit ex $\frac{1}{2}$𝟑 $+ \frac{1}{2}$𝒆 per 2: fit $\frac{2}{2}$𝟑 $+ \frac{2}{2}$𝒆, unde sublata 1𝒆 fit 1𝟑; item per 3 ex producto tollendo 2𝒆 fit pondus pentagonalium, etc.

20 Ita etiam polygonales regulariter fieri debent:

18 Solids are best of all formed by superimposing gnomons having always one angle vacant, and from which the whole figure can be resolved into triangles. From which it is easily recognized that the weights of all the polygonals are obtained by multiplying the triangulars by the numbers 2, 3, 4, 5, 6, etc., and subtracting from the product 1, 2, 3, 4 radices, etc.

19 So to form the tetragonal weight, $\frac{1}{2}n^2 + \frac{1}{2}n$ [multiplied] by 2 gives $\frac{2}{2}n^2 + \frac{2}{2}n$, from which subtracting n gives n^2; in the same manner by 3, and subtracting $2n$ from the product, gives the pentagonal weight, etc.

20 In this manner the polygonals can be regularly formed:

Triangular		Cubic		Pentagonal		Hexagonal	
R–A,	0	R–A,	0	R–A,	0	R–A,	0
1–0 ,	1	2–1 ,	1	3–2 ,	1	4–3 ,	1
2–0 ,	3	4–1 ,	4	6–2 ,	5	8–3 ,	6
3–0 ,	6	6–1 ,	9	9–2 ,	12	12–3 ,	15
4–0 ,	10	8–1 ,	16	12–2 ,	22	16–3 ,	28

Comments. The first sentence of the first paragraph is obscure: the notes in the *Oeuvres* call it "very elliptical." Foucher de Careil omitted some words he could not make out, while Prouhet ignored it completely. De Jonquières' reconstruction added thirty words and his translation further expanded it to over eighty words; this was in an attempt to explain a transition from the solids of the opening phrase to the plane figures of the rest of the sentence and paragraph.

The sentence is correctly rendered according to the manuscript but presents a difficulty in meaning. This is due to the presence of the words "solida" (solids) which makes the opening phrase inconsistent with the rest of the sentence and paragraph, which concern triangles and polygons. The difficulty disappears if it is assumed that "solida" is a copying error for "plana" (planes) in the original manuscript, and this position is taken here for the purpose of simplifying the explanation. The use of the word "plane," for plane figure or polygon (as "solid" was used for solid figure or polyhedron) was not uncommon, particularly when the interior of the polygon was also contemplated. Thus Kepler, like Pappus, uses "plane" to denote the polygonal face of a polyhedron. Descartes uses "plane" in Paragraph 21 and "surface" in Paragraph 22 for "plane figure."

The three paragraphs considered here, dealing with polygonal numbers, serve as an introduction for the extension of the concepts to space figures, polyhedra. They are noted under two heads: (a) gnomons and (b) division into triangles.

(a) *Gnomons.* Consider the first part of the first sentence of Paragraph 18, and the sentence of Paragraph 20, which are rewritten as:

(18) Plane figures (polygons) are formed by superimposing gnomons each of which has one angle vacant.

(20) In this manner the polygonal numbers can be regularly formed.

The first sentence is taken as referring to the polygons as such, from which, particularly the square, the concept of a gnomon presumably originated and was extended to the corresponding polygonal numbers. This has been explained above in Section 10 (p. 87) under the heading "Gnomons and division into triangles." Figure 16 (p. 88) shows the successive gnomons for the pentagonal numbers, and is referred to again here.

The four tables in Paragraph 20 show the formation by gnomons of successive triangular, square, pentagonal and hexagonal numbers, respectively. The third one is repeated here with added material:

	Pentagonals	
n	R − A,	0
	3 − 2,	1
	6 − 2,	5
	9 − 2,	12
4	12 − 2,	22
n	$3n - 2$,	$p(5,n)$

Descartes lists, in three columns, the number of sides of the gnomon (the 'radices', denoted by 'R'), the number of corners of the gnomon (the 'anguli,' denoted by 'A'), and the number of dots in the corresponding figurate number (see below for the explanation of why this column is headed by a zero). To this we have prefixed a column giving n, the number of dots along a side, or the order of the number.

As can be seen from the figure, the gnomon for the pentagonal numbers has three sides and two corners (angles). The number of dots to be added to the $(n-1)$th number to give the nth number is calculated by first considering the three sides of the gnomon separately, giving $3n$ dots, and then subtracting 2 for the two corners, whose dots have been counted twice. The last column of the table gives the pentagonal numbers themselves. Look at the line for $n = 4$: the gnomon has 3 times 4, minus 2 dots: the resulting number, 10, is added to the preceding (third) pentagonal number, 12, giving the fourth one, 22. The calculation of the gnomons is carried back to $n = 1$ (or rather it starts with $n = 1$), where the formula $3n - 2$ gives 1 for the gnomon, which, added to the zero of the preceding line, gives 1 for the first pentagonal number.

The other three tables follow the same routine. In general the recursion formula can be written

$$[(a - 2)n - (a - 3)] + p(a, n - 1) = p(a, n),$$

where a is the number of sides of the polygon and $p(a,n)$ stands for the nth a-gonal number.

(b) *Division into triangles.* The second part of the first sentence of Paragraph 18 indicates that polygons can be divided into triangles from the gnomons. The relation of gnomons to this division into triangles has been indicated in Section 10, p. 89 (see particularly Fig. 20). The next sentence and Paragraph 19 deal with the division into triangles.

Descartes gives a general method and two examples for finding the "weight" (general formula) for polygonal numbers. The following explanation uses the second example, the pentagon, and refers to Fig. 18 (p. 89). Lines drawn from one angle of the pentagon divide it into three triangles as shown in Fig. 18(a); this is as many triangles as there are sides in the gnomon. Pairs of these triangles have a common side; there are as many common sides, 2, as there are corners in the gnomon. The three triangles are separated in Fig. 18(b), which repeats the two common sides. The number of dots in the particular figure is calculated by multiplying the number of dots in one triangle by 3, and then subtracting twice the number of dots along a side, to allow for the 2 sides which have been counted twice. Stated generally for pentagons, this is three times the nth triangular number minus $2n$.

Descartes' general formula is thus seen to be

$$p(a, n) = (a - 2) \cdot p(3, n) - (a - 3)n,$$

where a is the number of sides of the polygon, $p(3,n)$ is the nth triangular number and $p(a,n)$ is the nth a-gonal number.

The weights, or general formulas, for various series of polygonal numbers are used by Descartes in deriving the weights of various polyhedral numbers. These are listed here (with the last one added) in the form in which he would have derived them in accordance with his method. Cf. p. 86.

$$p(3,n) = p(3,n) = (n^2 + n)/2$$
$$p(4,n) = 2p(3,n) - n = n^2$$
$$p(5,n) = 3p(3,n) - 2n = (3n^2 - n)/2$$
$$p(6,n) = 4p(3,n) - 3n = 2n^2 - n$$
$$p(8,n) = 6p(3,n) - 5n = 3n^2 - 2n$$
$$p(10,n) = 8p(3,n) - 7n = 4n^2 - 3n$$
$$p(12,n) = 10p(3,n) - 9n = 5n^2 - 4n.$$

21 Quod si imaginaremur figuras istas ut mensurabiles, tunc unitates omnes intelligerentur esse eiusdem rationis ac figurae ipsae: nempe in triangulis unitates triangulares; pentagona metiuntur per unitatem pentagonam etc. Tunc eadem esset proportio plani ad radicem quae est quadrati ad suam radicem; et solidi quae est cubi: ut si radix sit 3, planum erit 9, solidum 27, etc., v.g. Quod etiam valet in circulo et sphaera aliisque omnibus. Si enim unius circuli circumferentia sit triplo maior altera, eiusdem aream continebit novies. Unde animadvertis has progressiones nostrae matheseos, *ℛ*, *ℨ*, *œ*, etc., non esse alligatas figuris lineae, quadrati, cubi, sed generaliter per illas diversas mensurae species designari.

21 If one considers these figures as measurable, then all the units are understood as being of the same kind as the figures themselves: that is for triangles a triangular unit; pentagons are measured by a pentagonal unit, etc. Then the proportion between a plane and its radix is the same as the square to its radix; and a solid as a cube: so if the radix is 3, the plane will be 9, the solid 27, etc., for example. This holds also for the circle and the sphere, and all other figures. For if the circumference of a circle is three times larger than another, the area of the first will be nine times larger. From which it is observed that our mathematical progression, n, n^2, n^3, etc., is not attached to linear, square, cubic, figures, but is designated generally by the diverse species of measure.

Comments. This paragraph, dealing with units and measures, seems unrelated to the rest of the manuscript, or, at best, the relationship is rather tenuous. It will not be discussed here as this would lead too far afield.[110]

A note in the *Oeuvres* (X, p. 688) states that the two figures reproduced above "correspond to triangular and pentagonal numbers." But this is evidently not the case in view of their location and form. The figure on the left appears to be an illustration of the use of triangular units; it is taken as representing a triangle with 3 linear units on each side (the radix), and divided into triangular units, of which there are 9, as there should be in accordance with the first two sentences of the paragraph. The pentagon on the right, however, shows that this cannot be conveniently done in the case of pentagonal units. Perhaps these figures were not in the original manuscript but were added by Leibniz in a hurried attempt to illustrate what he had just copied.

22 Quinque corpora regularia, simpliciter ut per se spectantur, formantur per additamentum gnomonis, ut superficies fuerant formatae:

22 The five regular bodies, considered simply *per se,* are formed by adding gnomons, as the surfaces were formed.

Tetrahedrons		Octahedrons		Icosahedrons	
F–R+A,	0	F– R+A,	0	F– R+A,	0
1 –0 + 0,	1	4 – 4 + 1,	1	15 –20 + 6,	1
3 –0 + 0,	4	12– 8 + 1,	6	45 –40 + 6,	12
6 –0 + 0,	10	24 –12 + 1,	19	90 –60 + 6,	48
10 –0 + 0,	20	40 –16 + 1,	44	150 –80 + 6,	124

Cubes		Dodecahedrons	
F– R+A,	0	F– R+ A,	0
3 – 3 + 1,	1	9 –18 + 10,	1
12 – 6 + 1,	8	45 –36 + 10,	20
27 – 9 + 1,	27	108 –54 + 10,	84
48 –12 + 1,	64	198 –72 + 10,	220.

Comments. This paragraph[111] and its tables are concerned with the five regular solids. The phrase "considered simply *per se"* may have been added to exclude the semiregular solids, which come later and involve some additional considerations.

The characteristics of the regular polyhedra which are utilized are (1) the faces are congruent regular polygons, and (2) all the solid angles are identical.

Gnomons. The five series of polyhedral numbers are formed from gnomons in a manner analogous to that in which the polygonal numbers were formed (see the comment on Paragraphs 18, 19, 20 under the heading *Gnomons,* p. 93). No explanation is given by Descartes, so we elaborate the method here, using the dodecahedron as an example. In so doing, we anticipate some terminology which will be explained later.

In a way analogous to the treatment of polygonal numbers, a polyhedral number is considered as a nested sequence of polyhedral surfaces having one corner in common. The outermost surface, say the nth, excluding the common corner and its incident lines and faces, is the gnomon which is added to the next lowest, $(n-1)$th, polyhedral number to produce the nth one. Figure 21 is a representation of the surface of the regular dodecahedron. It has 12 pentagonal faces, 20 corners (triple points, points serving as the vertices of 3 plane angles), and 30 sides (double sides, lines serving as the sides of two faces). If one corner and its incident lines and faces are removed, the surface loses 3 faces, 12 double sides (3 are removed entirely and 9 are reduced to single sides), and 10 triple points (one is removed entirely, 6 are reduced to double points and 3 are reduced to single points). The result, the surface with one vacant angle, is the gnomon for the dodecahedral numbers; it has 9 pentahedral faces, 18 double sides, and 10 corners (triple points); only these elements are to be considered.

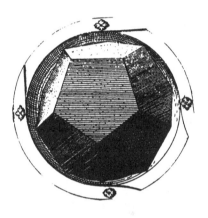 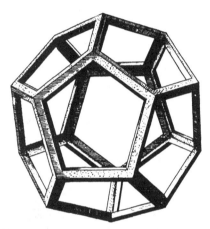

Figure 21

The utilization of the gnomons to form the dodecahedral numbers is illustrated by the fifth table, which is repeated here with some added material.[112]

	Dodecahedron	
n	$F_5 - R + A,$	0
1	$9 - 18 + 10,$	1
2	$45 - 36 + 10,$	20
3	$108 - 54 + 10,$	84
4	$198 - 72 + 10,$	220
n	$9p(5,n) - 18n + 10,$	

First, the faces of the nth gnomon are considered separately; each one is a pentagonal number, and hence the number of dots is 9 times the nth pentagonal number. But the faces have 18 common sides and the number of dots in these double sides (which in each is n) have been counted twice, hence $18n$ must be subtracted. Since 3 double sides meet at a corner (triple point), of which there are 10, there have been 3 subtractions of each of these points (eliminating the dots at the corners), and hence 10 must be added to compensate for the over-subtraction. The result is the value of the nth gnomon, which is added to the preceding, $(n-1)$th, dodecahedral number to form the nth one.

For a specific example consider the line of the table for $n = 4$. The 4th pentagonal number is 22, hence we get 198 for the 9 pentagons; 4 times 18 is subtracted and 10 then added. The result, 136, is the value of the 4th gnomon which, when added to third dodecahedral number, 84, gives the fourth one, 220.

The same procedure is applied separately to each of the other four regular polyhedra to obtain the other tables. In the case of the icosahedron (see Fig. 22), which has 20 triangular faces, 30 double sides, and 12 corners, the gnomon has 15 triangular faces, 20 double sides and 6 corners, giving the formula $15p(3,n) - 20n + 6$ for the nth gnomon. For the octahedron (see Fig. 23), which has 8 triangular faces, 12 double sides and 6 corners, the gnomon has 4 triangular faces, 4 double sides and only one corner. The gnomon for the cube has 3 square faces, 3 double sides and one corner, while for the tetrahedron the gnomon is only a single triangular face.

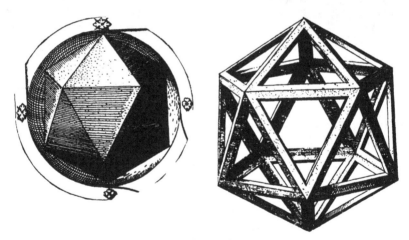

Figure 22

Division into pyramids, weights. The summary table at the end of the manuscript gives the general formula, called the weight, for each of these five series of polyhedral numbers. The method for obtaining them is not described, but the method for obtaining the formulas for polyhedral numbers corresponding to Archimedean solids is described in Paragraphs 30, 31 and 32, generally and with a specific example. This method, as applied to the regular solids, will be used here.

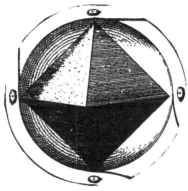

Figure 23

The polyhedron is divided into pyramids, as the polygons are divided into triangles, to obtain their weights. It is hence necessary to have the formulas (weights) for each series of pyramidal numbers. Descartes' method of obtaining these weights, as described in Paragraphs 30, 31 and 32, is done in two steps:

(1) add $n/2$ to the formula for the polygon of the base;
(2) multiply the result by $(n + 1)/3$.

Stated as a formula, this is

$$P(a,n) = (p(a,n) + n/2) \cdot (n + 1)/3,$$

where a is the number of sides of the base. How Descartes arrived at this expression for the general formula for a pyramidal number is not evident.

The pyramidal formulas which would be needed in connection with the 14 polyhedra mentioned in the manuscript are listed in the following table.

$P(3,n)$	$=$	$(n^2 + 2n)$	$(n + 1)/6$	$=$	$(n^3 + 3n^2 + 2n)/6$
$P(4,n)$	$=$	$(2n^2 + n)$	$(n + 1)/6$	$=$	$(2n^3 + 3n^2 + n)/6$
$P(5,n)$	$=$	$(3n^2)$	$(n + 1)/6$	$=$	$(3n^3 + 3n^2)/6$
$P(6,n)$	$=$	$(4n^2 - n)$	$(n + 1)/6$	$=$	$(4n^3 + 3n^2 - n)/6$
$P(8,n)$	$=$	$(6n^2 - 3n)$	$(n + 1)/6$	$=$	$(6n^3 + 3n^2 - 3n)/6$
$P(10,n)$	$=$	$(8n^2 - 5n)$	$(n + 1)/6$	$=$	$(8n^3 + 3n^2 - 5n)/6$
$P(12,n)$	$=$	$(10n^2 - 7n)$	$(n + 1)/6$	$=$	$(10n^3 + 3n^2 - 7n)/6$

Only the first and third of these formulas are stated explicitly in the manuscript.

Now for the polyhedron. Lines are drawn from the common corner of the nested set of polyhedral surfaces to each of the corners with which it is not already connected by a line; pairs of these lines will stand joined by a line on the surface joining their extremities, and a triangular plane surface is imagined to be connecting each of these triplets. The result is a set of pyramids having a common apex, pairs of which have a common lateral triangular face. The base of each pyramid is a face of the gnomon.

Now consider the dodecahedron and the derivation of its gnomon as given above. When divided into pyramids, there will be as many pyramids as there are faces in the gnomon, each of these faces being the base of one of the pyramids. Since the gnomon has 9 pentagonal faces there will be 9 pentagonal pyramids. Each of the double lines of the gnomon marks a common face of two pyramids; hence there are 18 common triangular faces corresponding to the 18 double sides of the gnomon. Each of the corners (triple points) of the gnomon, of which there are 10, marks the end of a line which is common to three faces.

The derivation of the formula (weight) follows. Consider the pyramids separately: the total number of dots in these is 9 times the number in each pyramid (which we denote by $P(5,n)$), or $9P(5,n)$. But the dots in 18 triangular faces have been counted twice, so $18p(3,n)$ dots are subtracted. But the dots in 10 lines have been subtracted three times (reducing them to zero), so the number of dots in 10 lines, $10n$, must be restored. This is expressed as

$$\text{Dodec.} \ = \ 9P(5,n) - 18p(3,n) + 10n,$$

and substitution of the formulas for $P(5,n)$ and $p(3,n)$ results in

$$\text{Dodec.} \ = \ (9n^3 - 9n^2 + 2n)/2.$$

The same procedure is followed for the other regular polyhedra, but for the tetrahedron there is only one pyramid, itself, and in the table for the cubic numbers it is obviously included merely to show the generality of the procedure.

Another method of deriving the formulas would be to assume a cubic polynomial with undetermined coefficients and then determine the coefficients from the first three or four of numbers in the tables.

23 Corporis quod constat 4 hexagonis et 4 triangulis, latera sunt 18, anguli 12, facies 8. Igitur huius gnomon constat 2 hexagonis et tribus triangulis faciebus, minus sex radicibus, + 2 angulis:

23a Horum autem differentias ita definiemus, prioris 1,

23 The body which is composed of 4 hexagons and 4 triangles has 18 sides, 12 angles, 8 faces. Hence its gnomon is composed of 2 hexagonal and 3 triangular faces, minus 6 radices, plus 2 angles:

	F +	F -	R +	A,	
Gnomon	3 +	2 -	6 +	2,	1
	9 +	12 -	12 +	2,	12
	18 +	30 -	18 +	2,	44
	30 +	56 -	24 +	2,	108
	45 +	90 -	30 +	2,	215

The top of the table header also shows: F + F − R + A,   0

23a Of these now the differences are thus defined, first 1,

1

11 ⌐ 10

32 ⌐ 21

64 ⌐ 32

107 ⌐ 43

161 ⌐ 54.

Comments. The present paragraph begins the consideration of the Archimedean or semiregular solids, which occupies the rest of the manuscript. These differ from the regular solids in that the faces are not all of the same type, but there may be two or three different types of faces, all regular polygons, in the same solid. They are similar to the regular solids in that all the solid angles are identical and each is identically related to the others. They were discovered by Archimedes but in extant literature are first described by Pappus.[113] The names used by Kepler[114] are still generally accepted, although alternative names for a few are also found.

The solid of the paragraph is the truncated tetrahedron. See Fig. 24. As stated in the text, it has 4 hexagonal and 4 triangular faces, 18 sides, and 12 angles. The word sides refers to the sides of the faces and each one serves two

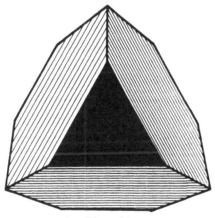

Figure 24

faces; since they are split into two and counted twice in the procedure, they are referred to as "double sides" in our discussion. The unmodified word "angles" is used throughout the descriptions; the Latin word "angulus" in general means "corner,"[115] and this word is used in our discussion instead of "solid angle."

If one corner and its attendant 3 faces are removed, we are left with a surface having 3 triangular and 2 hexagonal faces for the gnomon. Three double sides are removed, and 9 double sides reduced to single sides, leaving 6 double sides for the gnomon. One corner is removed and 9 reduced, leaving 2 for the gnomon. The calculation of the polyhedral numbers from the gnomons is shown by the table, which is repeated here with some additions.

n	$F_3 +$	$F_6 -$	$R +$	$A,$	0	d_1	d_2	d_3
1	3 +	2 −	6 +	2,	1	1		
2	9 +	12 −	12 +	2,	12	11 — 10		
3	18 +	30 −	18 +	2,	44	32 ⌇ 21	11	
4	30 +	56 −	24 +	2,	108	64 ⌇ 32	11	
5	45 +	90 −	30 +	2,	215	107 ⌇ 43	11	
6	63 +	132 −	36 +	2,	376	161 ⌇ 54	11	

F_3 refers to the triangular faces and F_6 to the hexagonal faces. Note that the first line of numbers gives the elements of the gnomon. If we consider the faces separately, we get $3p(3,n)$ dots from the 3 triangles and $2p(6,n)$ from the 2 hexagons; these are added together. The excess dots from the 6 sides which have been counted twice, $6n$, are subtracted, and 2 dots for the over-subtraction at the 2 corners are then added. Hence the formula for the nth gnomon is

$$3p(3,n) + 2p(6,n) - 6n + 2.$$

The procedure and the significance of the table are the same as those described for the dodecahedron at Paragraph 22 (p. 100), except that there are two kinds of faces in the gnomon.

If the truncated tetrahedron is divided into pyramids, there are 3 triangular and 2 hexagonal pyramids, 6 common triangular faces, and 2 lines common to three pyramids. The formula then is

$$\begin{aligned} \text{Trun. Tetra.} &= 3P(3,n) + 2P(6,n) - 6p(3,n) + 2n \\ &= (11n^3 - 3n^2 - 2n)/6. \end{aligned}$$

24 Corporis quod constat 8 triangulis et 6 quadratis faciebus, latera sunt 24, anguli 12 et facies 14. Et huius gnomon constat 6 triangulis et 4 quadratis faciebus, − 14 radicibus, + 5 angulis:

24 The body which is composed of 8 triangular and 6 square faces, has 24 sides, 12 angles and 14 faces. Its gnomon is composed of 6 triangular and 4 square faces, minus 14 radices, plus 5 angles:

	F +	F −	R +	A,	0
Gnomon	6 +	4 −	14 +	5,	1
	18 +	16 −	28 +	5,	12
	36 +	36 −	42 +	5,	47
	60 +	64 −	56 +	5,	120
					(245).

Comments. The polyhedron of this paragraph is the cuboctahedron (see Fig. 25). The formula for the *n*th gnomon is seen at once to be

$$6p(3,n) + 4p(4,n) - 14n + 5.$$

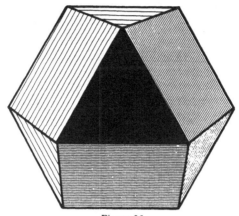

Figure 25

And the formula (weight) of the polyhedral number can also be written down in the form

$$6P(3,n) + 4P(4,n) - 14p(3,n) + 5n,$$

which gives

$$\text{Cuboct.} = (7n^3 - 6n^2 + 2n)/3.$$

25 Corporis quod constat 8 hexagonis et 6 quadratis faciebus, latera sunt 36, anguli 24 et facies 14. Huius gnomon habet 6 hexagonas et 5 quadratas facies, minus 23 radices, + 13 angulos:

25 The body which is composed of 8 hexagonal and 6 square faces, has 36 sides, 24 angles and 14 faces. Its gnomon has 6 hexagonal and 5 square faces, minus 23 radices, plus 13 angles:

	$F + F - R + A,$	0
Gnomon	$6 + 5 - 23 + 13,$	1
	$36 + 20 - 46 + 13,$	24
	$90 + 45 - 69 + 13,$	103
	$168 + 80 - 92 + 13,$	$272.$

Comments. The polyhedron of this paragraph is the truncated octahedron (see Fig. 26). The formula for its gnomon is

$$6p(6,n) + 5p(4,n) - 23n + 13,$$

and for the polyhedral number

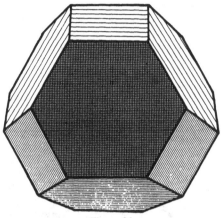

Figure 26

Trunc. Oct.   = $6P(6,n) + 5P(4,n) - 23p(3,n) + 13n$
              = $(17n^3 - 18n^2 + 4n)/3.$

26 Corporis quod constat 8 triangulis et 6 octangulis faciebus, latera 36, anguli 24, facies 14. Huius gnomon habet 4 octagonas et 7 triangulares facies, minus radices 20, plus angulos 10:

26 The body which is composed of 8 triangular and 6 octagonal faces, [has] sides 36, angles 24, faces 14. Its gnomon has 4 octagonal and 7 triangular faces, minus 20 radices, plus 10 angles:

	$F +$	$F - R + A,$	0
Gnomon	$7 +$	$4 - 20 + 10,$	1
	$21 +$	$32 - 40 + 10,$	24
	$42 +$	$84 - 60 + 10,$	100
	$70 +$	$160 - 80 + 10,$	260.

Comments. This polyhedron is the truncated cube (see Fig. 27). The formula for the gnomon is

$$7p(3,n) + 4p(8,n) - 20n + 10,$$

and for the polyhedral number

Trunc. Cube   = $7P(3,n) + 4P(8,n) - 20p(3,n) + 10n$
              = $(31n^3 - 27n^2 + 2n)/6.$

27 Corporis quod constat 18 quadratis et 8 triangulis, latera sunt 48 et anguli 24 et facies 26. Huius autem gnomon constat 15 quadratis et 7 triangulis faciebus, − 37 radicibus, plus 16 angulis:

27 The body which is composed of 18 squares and 8 triangles, has 48 sides and 24 angles and 26 faces. Its gnomon is composed of 15 square and 7 triangular faces, minus 37 radices, plus 16 angles:

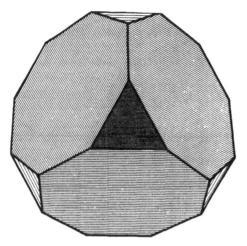

Figure 27

	$F+$	$F-$	$R+$	$A,$	0
Gnomon	$7+$	$15-$	$37+$	$16,$	1
	$21+$	$60-$	$74+$	$16,$	24
	$42+$	$135-$	$111+$	$16,$	106
	$70+$	$240-$	$184+$	$16,$	$284.$

Comments. This polyhedron is the rhombicuboctahedron (see Fig. 28). The formula for its gnomon is

$$7p(3,n) + 15p(4,n) - 37n + 16,$$

and for the figurate polyhedral number

$$7P(3,n) + 15P(4,n) - 37p(3,n) + 16n$$

or

$$(37n^3 - 45n^2 + 14n)/6.$$

28 Corporis quod constat 12 pentagonis et 20 hexagonis faciebus, latera sunt 90, anguli 60 et facies 32. Huius gnomon habet 11 pentagonas et 18 hexagonas facies, minus 76 radices, plus 48 angulos:

28a Qui ad sinistrum latus lineae characteres in Mso elisi et dubii erant. (Neque hic gnomon cum numeris convenit ut in prioribus.)

28 The body which is composed of 12 pentagonal and 20 hexagonal faces, has 90 sides, 60 angles and 32 faces. Its gnomon has 11 pentagonal and 18 hexagonal faces, minus 76 radices, plus 48 angles.

	$F+$	$F-$	$R+A,$		0
Gnomon	$11+$	$18-$	$76+48,$		1
	$55+108-$		$152+48,$		60
	$132+270-$		$228+48,$		282

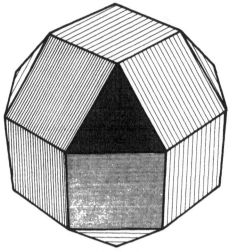

Figure 28

28a The characters on the left side of the line in the manuscript were washed out and doubtful. (Nor does the gnomon agree with the numbers as in the preceding).

Comments. This polyhedron is the truncated icosahedron (see Fig. 29). The formula for its gnomon is

$$11p(5,n) + 18p(6,n) - 76n + 48,$$

and for the figurate number

$$11P(5,n) + 18P(6,n) - 76p(3,n) + 48n$$

or

$$(35n^3 - 47n^2 + 14n)/2.$$

As stated in Section 3 (p. 9) the manuscript contains no descriptive section corresponding to this table. Such a description has accordingly been supplied, on the model of the preceding and following paragraphs. Paragraph 28a is a marginal note, obviously by Leibniz. It indicates that there was some difficulty and confusion in the original manuscript.

29 Corpus ex 20 triangulis et 12 pentagonis: latera 60, anguli 30, facies 32, et huius gnomon habet 18 triangulas et 10 pentagonas facies, minus radices 48, plus 21 angulis:

29 The body formed of 20 triangles and 12 pentagons [has] 60 sides, 30 angles, 32 faces, and its gnomon has 18 triangular and 10 pentagonal faces, minus 48 radices, plus 21 angles:

	$F+$	$F-$	$R+$	$A,$	0
Gnomon	$18+$	$10-$	$48+$	$21,$	1
	$54+$	$50-$	$96+$	$21,$	30
	$108+$	$120-$	$144+$	$21,$	$135.$

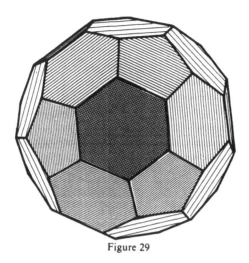

Figure 29

Comments. This polyhedron is the icosidodecahedron (see Fig. 30). The formula for its gnomon is

$$18p(3,n) + 10p(5,n) - 48n + 21$$

and for the figurate number,

$$18P(3,n) + 10P(5,n) - 48p(3,n) + 21n$$

or

$$(8n^3 - 10n^2 + 3n).$$

The table of this paragraph has been misplaced in the manuscript, where it precedes Paragraph 23. It was restored to its proper place by Prouhet II and de Jonquières.

The formula (weight) for this polyhedron is worked out in detail by Descartes in Paragraphs 31 and 32, as an example of his method of deriving the weights.

30 Termini algebraici aequales istis numeris figuratis inveniuntur ducendo exponentem faciei $+ \frac{1}{2}\mathcal{Z}$ per $\frac{1}{2}\mathcal{Z}+\frac{1}{2}$, deinde per numerum facierum; hocque toties faciendo, quot sunt diversa genera facierum in dato corpore; deinde producto addendo vel tollendo numerum radicum ductum per $\frac{1}{2}\mathcal{Z} + \frac{1}{2}\mathcal{Z}$, et numerum angulorum ductum per $1\mathcal{Z}$.

31 Ut si quaerantur termini adaequales numeris figuratis qui repraesentent corpus ex 20 triangulis et 12 pentagonis, quoniam gnomon huius corporis constat 18 triangularibus faciebus et 10 pentagonis, minus 48 radicibus, $+ 21$ angulis, primo addo $\frac{1}{2}\mathcal{Z}$ numero $\frac{1}{2}\mathcal{Z} + \frac{1}{2}\mathcal{Z}$, qui est exponens faciei triangularis, et productum, nempe $\frac{1}{2}\mathcal{Z} + 1\mathcal{Z}$, duco per $\frac{1}{2}\mathcal{Z}+\frac{1}{2}$: fit $\frac{1}{6}\mathcal{C}+\frac{1}{6}\mathcal{Z}+\frac{1}{6}\mathcal{Z}$, quod duco per 18 et fit $3\mathcal{C} + 9\mathcal{Z} + 6\mathcal{Z}$.

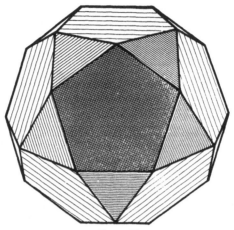

Figure 30

32 Deinde addo etiam $\frac{1}{2}$ 𝓍 numero $\frac{3}{2}$ 𝓎 − $\frac{1}{2}$𝓍, qui est exponens faciei pentagonalis, et fit$\frac{3}{2}$ 𝓎, quo ducto per$\frac{1}{3}$ 𝓍+$\frac{1}{3}$, fit$\frac{1}{2}$ 𝓪+$\frac{1}{2}$𝓎; et deinde per 10, fit 5 𝓪 + 5 𝓎; quod si iungatur cum numero praecedenti, fit 8 𝓪 + 14 𝓎 + 6𝓍. Unde si tollatur numerus radicum 48 ductus per $\frac{1}{2}$ 𝓎 + $\frac{1}{2}$𝓍 nempe 24 𝓎 + 24𝓍, fit 8 𝓪 − 10 𝓎 − 18𝓍; cui si addatur 21𝓍 propter 21 angulos, fit 8𝓪 − 10 𝓎 + 3𝓍, numerus algebraicus quaesitus.

30 The algebraic expressions for these figurate numbers are found by multiplying the exponent of the face plus$\frac{1}{2}n$ by$\frac{1}{3}n + \frac{1}{3}$, then by the number of faces, which is done as many times as there are different types of faces in the given body; then add to or subtract from the result the number of radices multiplied by$\frac{1}{2}n^2 + \frac{1}{2}n$, and the number of angles multiplied by n.

31 So if the expression for the figurate numbers which represent the body [consisting] of 20 triangles and 12 pentagons is sought, since the gnomon of this body is composed of 18 triangular and 10 pentagonal faces, less 48 radices, plus 21 angles, I first add$\frac{1}{2}n$ to the number$\frac{1}{2}n^2 + \frac{1}{2}n$, which is the exponent of a triangular face, and multiply the result, namely$\frac{1}{2}n^2 + n$, by$\frac{1}{3}n + \frac{1}{3}$: this gives$\frac{1}{6}n^3 + \frac{3}{6}n^2 + \frac{2}{6}n$, which I multiply by 18, which gives $3n^3 + 9n^2 + 6n$.

32 Then I also add$\frac{1}{2}n$ to the number$\frac{3}{2}n^2 − \frac{1}{2}n$, which is the exponent of the pentagonal face, giving$\frac{3}{2}n^2$, which multiplied by$\frac{1}{3}n + \frac{1}{3}$gives$\frac{1}{2}n^3 + \frac{1}{2}n^2$; and then by 10, giving $5n^3 + 5n^2$, which is joined to the preceding number, giving $8n^3 + 14n^2 + 6n$. From which is subtracted the number of radices, 48, multiplied by $\frac{1}{2}n^2 + \frac{1}{2}n$, that is, $24n^2 + 24n$, giving $8n^3 − 10n^2 − 18n$, to which is added $21n$ for the 21 angles, giving $8n^3 − 10n^2 + 3n$, the algebraic number sought.

Comments. These paragraphs give Descartes' method of deriving the formulas (weights) for the polyhedral figurate numbers. Paragraph 30 states the method generally and the next two paragraphs work out an example. Note that the term "exponent" is used for the formula of polygonal numbers. The method has been described and applied in preceding paragraphs and is summarized here to show the source.

On the basis of all three paragraphs, the method may be outlined as follows.

1. Determine the pyramidal number corresponding to one type of face in the gnomon. The pyramidal number is found by adding $n/2$ to the formula (exponent) of the polygonal number corresponding to the face, and the result then multiplied by $(n + 1)/3$. See under Paragraph 22, p. 99.

2. Multiply the result of step 1 by the number of such faces in the gnomon.

3. Perform steps 1 and 2 for each other type of face in the gnomon.

4. Add together the results of the preceding steps.

5. Subtract from the result of step 4 the product of the number of radices (double sides) in the gnomon by $(n^2 + n)/2$ (the formula for a triangular number). This is to remove double counts of adjoining triangular faces of the pyramids.

6. Add, to the result of step 5, as many radices as there are corners (angles) in the gnomon. This is to compensate for the over-subtraction in step 5.

The example stated in the text, which is the polyhedron of Paragraph 29, can readily be followed from the above outline.

33 Denique pondera omnium 14 solidorum prout imaginamur illa oriri ex progressionibus arithmeticis:

(Large table)

33a $\sqrt{\frac{y}{2} - 6\sqrt{2}}$
  nescio cur

34 (Alio atramento ascriptum erat)
Supersunt duo corpora, unum ex 6 octogonis, 8 hexagonis et 12 quadratis, aliud ex 30 quadratis, 12 decag. et 20 hexag.

33 Finally here are the weights of all the 14 solids, as we imagine them to result from arithmetical progressions:

(Large table; see pp. 110-111)

33a $\sqrt{\frac{y}{2} - 6\sqrt{2}}$
I know not why

34 (Inserted in another [color of] ink). There remain two bodies, one composed of 6 octagons, 8 hexagons and 12 squares, the other of 30 squares, 12 decagons and 20 hexagons.

Comments. The table presented here is a little more than a translation of the table of the manuscript. The lines have been numbered, headings changed, the identifications of the polyhedra in column 1 amplified, and blanks filled in. Some of the arithmetical expressions in columns 3-5 have been simplified, thus $\sqrt{128}$ is written $8\sqrt{2}$, $\sqrt{3/4}$ is written $\sqrt{3}/2$, etc., and some of the more complicated expressions also have been simplified. The few arithmetical corrections to the manuscript have been noted on p. 28.

[1] Polyhedron	[2] Formula	[3] Volume	[4] Diameter	[5]	
1. Tetrahedron 4 triangles	$(n^3 + 3n^2 + 2n)/6$	$(\sqrt{2}/12)n^3$	$(\sqrt{3/2}\,)n$	from cube with side	$\sqrt{2}n/2$
2. Octahedron 8 triangles	$(2n^3 + n)/3$	$(\sqrt{2}/3)n^3$	$(\sqrt{2})n$	from tetrahedron with side	$2n$
3. Cube 6 squares	n^3	n^3	$(\sqrt{3})n$		
4. Icosahedron 20 triangles	$(5n^3 - 5n^2 + 2n)/2$	$(5\sqrt{5}/12 + 5/4)n^3$	$(\sqrt{5/2 + \sqrt{5}/2}\,)n$		
5. Dodecahedron 12 pentagons	$(9n^3 - 9n^2 + 2n)/2$	$(7\sqrt{5}/4 + 15/4)n^3$	$(\sqrt{15/2 + \sqrt{3}/2}\,)n$		
6. Truncated tetrahedron 4 triangles, 4 hexagons	$(11n^3 - 3n^2 - 2n)/6$	$(23\sqrt{2}/12)n^3$	$(\sqrt{11/2}\,)n$	from tetrahedron with side	$3n$
7. Cuboctahedron 8 triangles, 6 squares	$(7n^3 - 6n^2 + 2n)/3$	$(5\sqrt{2}/3)n^3$	$2n$	{from octahedron with [side] {from cube [with side]	$2n$ $\sqrt{2}n$
8. Truncated octahedron 6 squares, 8 hexagons	$(17n^3 - 18n^2 + 4n)/3$	$(8\sqrt{2})n^3$	$(\sqrt{10})n$	octahedron	$3n$

9. Truncated cube 8 triangles, 6 octagons	$(31n^3 - 27n^2 + 2n)/6$	$(14\sqrt{2}/3 + 7)n^3$	$(\sqrt{7+4\sqrt{2}}\,)n$	cube	$(\sqrt{2}+1)n$
10. Rhombicuboctahedron 8 triangles, 18 squares	$(37n^3 - 45n^2 + 14n)/6$	$(10\sqrt{2}/3 + 4)n^3$	$(\sqrt{5+2\sqrt{2}}\,)n$	$\begin{cases} \text{cube} \\ \text{octahedron} \end{cases}$	$(\sqrt{2}+1)n$ $(3\sqrt{2}+2)n/2$
11. Truncated icosahedron 12 pentagons, 20 hexagons	$(35n^3 - 47n^2 + 14n)/2$	$(43\sqrt{5}/4 + 125/4)n^3$	$(\sqrt{29/2+9\sqrt{5}/2}\,)n$	icosahedron	$3n$
12. Icosidodecahedron 20 triangles, 12 pentagons	$(8n^3 - 10n^2 + 3n)$	$(17\sqrt{5}/6 + 15/2)n^3$	$(1 + \sqrt{5})n$	$\begin{cases} \text{icosahedron} \\ \text{dodecahedron} \end{cases}$	$2n$ $(\sqrt{5}-1)n$
13. Truncated dodecahedron 20 triangles, 12 decagons	$(33n^3 - 41n^2 + 10n)/2$	$(235\sqrt{5}/12 + 165/4)n^3$	$(\sqrt{37/2+15\sqrt{5}/2}\,)n$	dodecahedron	$\sqrt{5}n$
14. Rhombicosidodecahedron 20 triangles, 30 squares, 12 pentagons	$(18n^3 - 25n^2 + 8n)$	$(29\sqrt{5}/3 + 20)n^3$	$(\sqrt{11+4\sqrt{5}}\,)n$	$\begin{cases} \text{icosahedron} \\ \text{dodecahedron} \end{cases}$	$(3\sqrt{5}-1)n/2$ $3(\sqrt{5}-1)n/2$

The second column of the table gives the formulas or weights of the series of polyhedral numbers corresponding to the identified polyhedra. Lines 1-5 relate to the regular polyhedra but are not listed in the same order as their tables in Paragraph 22. Lines 6-12 correspond to the polyhedra of Paragraphs 23 to 29, respectively. Lines 13 and 14 relate to polyhedra not previously mentioned in the manuscript; the formulas for these are not in the table in the manuscript; they were worked out by Prouhet and de Jonquières and added to the table in the *Oeuvres*. These two are the truncated dodecahedron and the rhombicosidodecahedron (see Figs. 31 and 32).

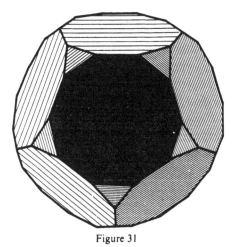

Figure 31

The table in the manuscript thus lists the five regular solids and nine of the semiregular ones. Since the sentence introducing the table refers to "all the 14 solids," it may be that descriptions and tables for 13 and 14 were in the original manuscript but were omitted by Leibniz in copying.

Two more Archimedean solids are referred to in the sentence below the table (Paragraph 34), making a total of eleven. These are the truncated cuboctahedron and the truncated icosidodecahedron (see Figs. 33 and 34).

Unaccountably, two of the thirteen Archimedean solids are not mentioned. They are the snub cube (32 triangles, 6 squares) and the snub dodecahedron (80 triangles, 12 pentagons); see Figs. 35 and 36. Perhaps Descartes thought he had put down enough for his preliminary notes. But it is worth remarking that the two he omitted are the only ones which cannot be formed by simple truncations of the regular solids.

The formulas for the polyhedral series corresponding to each of the four solids last mentioned can be derived readily from the data given.

The third column of the table, headed "geometrical weight" in the manuscript and "Volume" here, gives the volume of each polyhedron listed, as a solid, in terms of the radix (side) cubed.[116]

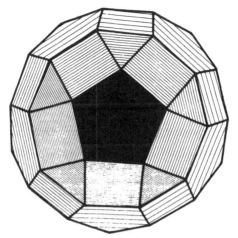

Figure 32

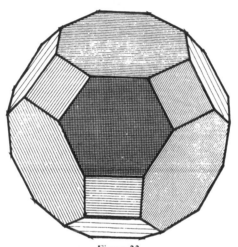

Figure 33

The fourth column, headed "major axes" in the manuscript, gives the diameters of the circumscribing spheres, in terms of the radix (side) of the polyhedron.[117]

The Archimedean solids listed can be derived from the regular solids by appropriate truncations (cutting off of corners). The fifth (last) column of the table gives the side of the original solid in terms of the side of one unit (radix) in its derived solid.[118]

The formula and comment of Paragraph 33a are an addition by Leibniz, with an indication (by a line drawn in the manuscript) that they refer to the

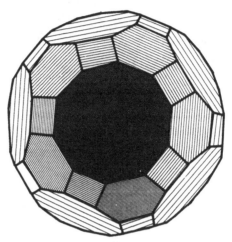

Figure 34

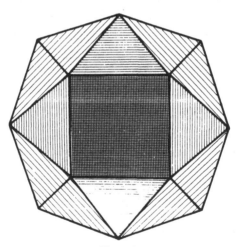

Figure 35

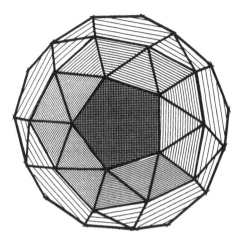

Figure 36

second formula in line 10 of the last column. [For an explanation of this see n. 15. G.J.T]

The last three columns of the table do not relate to polyhedral numbers or to Part I of the manuscript and introduce no new matters of principal. They are concerned with the same sort of things dealt with before Descartes' time, measurements and numerical expressions. Perhaps some of the numerical expressions given by Descartes were new, but no effort has been made to determine this.

12 General Comments

The first question to be considered is the indebtedness, if any, of Descartes to Faulhaber.

Prouhet refers to Faulhaber's *Numerus figuratus* of 1614 (which was not available to him), as reported by Kästner.[119] He states that (according to Kästner) Faulhaber considered polyhedral numbers and gave a table of nine columns of figurate numbers, of which six were polyhedral, with the cossic expression for each at the head of its column. Prouhet adds, "It is evident that Descartes borrowed the fundamental idea of his work from Faulhaber." Milhaud ("L'oeuvre de Descartes") followed Prouhet without having seen either Kästner or Faulhaber. From this account it would appear that all that Descartes did was to add eight more polyhedral numbers to the six shown by Faulhaber. Prouhet gives the wrong impression. Kästner's account of Faulhaber's work mentions only pyramidal numbers. In one place the pyramidal numbers are referred to as "corporales numeri"; Prouhet translates "corporales" as "polyèdres," although it would be more appropriately translated as "corporeal," and, without paying attention to the context, indicates that Faulhaber showed six polyhedral numbers, thus giving a completely false impression of Descartes' work.

Since Faulhaber's work preceded Descartes, and Descartes was acquainted with him and undoubtedly with his work (see above p. 30), some account will be given of it, derives in the first place from Kästner. In his section of "Coss and algebra," Kästner reviews a number of works of Faulhaber.

Johann Faulhaber (1580-1635) was a "rechenmeister und modist," a teacher of mathematics in Ulm. Besides what might be called regular works he wrote a number of tracts on figurate numbers—"tracts," because they are short and of a cabalistic nature. Some of the titles (abbreviated) are: "Neuer mathematischer Kunstspiegel, darinnen fürnehmlich dreyerley Stück zu Sehen," Ulm, 1612, 28 pages; "Andeutung einer unerhörten neuen Wunderkunst," Ulm, 1613; "Numerus figuratus," 1614, 24 pages; "Gemein offen

Ausschreiben...an alle Philosophos, Mathematicos, sonderlich Arithmeticos und Künstler," Augsburg, 1615, 4 pages; "Miracula Arithmetica," Augsburg, 1622, 93 pages; "Vernünftiger Creaturen Weissagungen," Augsburg, 1632.

Faulhaber treated polygonal and pyramidal numbers from the standpoint of arithmetical series, and gave some individual formulas for the numbers of sides. His main concern was to find arithmetical series which produced special numbers mentioned in the Bible or having some religious significance. An example: after referring to the 120 years of Genesis 6:3, he states: "The triangular number 120 is also a certain pyramidal number, which I investigated through a completely new invention."

Then he asks for the cubicossic equation (cubic polynomial) which such number must have from "my new invention." (Apparently Faulhaber's new invention was the method of finding pyramidal numbers.) In answer he gives the formula for the pyramidal number with base of 13 sides,

$$(11n^3 + 3n^2 - 8n)/6,$$

which gives 120 when $n = 4$. The number 120 is the 15th triangular number.

Another miraculous number (Wunderzahl) is 490, the 70 weeks (490 days) of Daniel 9:24–27, a pyramidal number which was divinely inspired. The formula of the pyramidal number with base of 50 sides is given by

$$(16n^3 + n^2 - 15n)/2.$$

This gives 490 when $n = 4$.

Another formula given is the pyramidal number with base of 154 sides,

$$(152n^3 + 3n^2 - 149n)/6.$$

When $n = 4$, this gives 1530, the date of the Augsburg Confession.

Some polygonal numbers mentioned by Kästner as being treated by Faulhaber are: 666, the number of the beast in Rev. 13:18, which is the 36th triangular number; 1335, the number of days in Daniel 12:12, which is the 30th pentagonal number; 2300, the number of days in Daniel 8:14, which is the 20th tetradecagonal number.

But Faulhaber may have had something new; what might now be called hyperpyramidal numbers. These are described briefly by Kästner. The pyramidal numbers are called corporeal numbers of the first kind; the successive summations of these give the corporeal numbers of the second kind, whose general term would be a polynomial of the fourth degree; these in turn are summed to form the corporeal numbers of the third kind, represented by a quintic equation; and so on, up to the sixth kind. Thus Faulhaber considered arithmetical series or progressions of orders higher than the 2 of triangular numbers and the 3 of pyramidal numbers.

The *Numerus figuratus* of Faulhaber[120] confirms Kästner's account, as summarized above. In addition, we note that it gives the cossic equations for the triangular, square, pentagonal and hexagonal, polygonal and pyramidal numbers. The triangular pyramidal numbers are summed to form the next higher stage of pyramids, etc., as has been stated. The results are shown by cubicossic equations and by a table.[121]

One cannot imagine Descartes being impressed by Faulhaber's cabalistic lore, nor is it likely that Descartes first learned about figurate numbers from him, as these were in the common stock of mathematical knowledge of the time.[122] However, he would no doubt have been interested in the algebra involved, and it is reasonable to infer that Faulhaber's enthusiasm may have inspired him to begin to think about the subject and to evolve some new ideas. Furthermore, Faulhaber's equations of pyramidal numbers may well have been the starting point for Descartes' treatment of pyramids. Descartes' polyhedral numbers were new and involved a new concept, although it was analogous to concepts of the Greeks.

The Greeks, as has been stated, only considered pyramidal numbers and the obvious cube; there was no concept of polyhedral numbers in general as there was of polygonal numbers.

However, it cannot be said that Descartes' results were entirely new at the time the manuscript was published in 1860. A book by Marpurg of 1774 gives the same formulas for the polyhedral numbers corresponding to the regular solids as did Descartes, and they were included in Klügel's mathematical dictionary in 1808.[123] Pollock in his 1850 paper referred to them generally. The numbers corresponding to the semiregular solids were evidently new in 1860, but apparently no one has paid any attention to them.

Comments on the second part of the manuscript were made by Foucher de Careil in his publication, then the same year by Prouhet I, later by de Jonquières and later still by Milhaud. Foucher de Careil refers to the manuscript in his introduction, considering it as one of the applications of the *Méthode,* with extravagant remarks. He does not specifically mention or discuss the first part of the manuscript, but as to the second part he refers to the new application of arithmetic to geometry, whereby all solids are represented as emerging from arithmetical progressions and submitted to calculation. Milhaud (1918) briefly discusses the manuscript but is mainly concerned with Part II in its relation to the problem of dating. He states that it appears that the geometrical first part was but a preface to the calculations which fill the second part.

I have found no other discussions of Part II, except for a few references to the fact that the second part of the manuscript relates to polyhedral numbers. Strangely enough, Dickson does not mention the manuscript, though he does refer to the letters to Mersenne of 1638 and to a small Descartes note relating to polygonal numbers.

Descartes' system of forming polyhedral numbers could not be carried much beyond the regular and semiregular solids. It could be applied to the series of prisms and antiprisms having the same properties as the Archimedean solids, but not to the third class of solids with all faces regular polygons (see n. 116) as in these the corners are not the same.

Except in Part II of the manuscript, which evidently was not carried further, Descartes does not appear to have done much with number theory. A theorem of Fermat is mentioned in two places. Fermat had sent a theorem to Mersenne, stating that he had a proof, but not revealing it. The theorem was that every number is either triangular or the sum of 2 or 3 triangular numbers; every number is either a square or the sum of up to 4 squares; pentagonal or the sum of up to 5 pentagonal numbers; and so on *ad infinitum*. Descartes mentions the theorem and says a few words about it in a letter to Mersenne of July 27, 1638 (*Oeuvres*, Vol. 2, pp. 253–280). After comments on a special case he says:

> But as for the theorem, which undoubtedly is one of the finest which can be found concerning numbers, I do not know the proof, and I believe it so difficult that I dare not undertake to seek it.

The same letter treats several problems involving polygonal numbers, as does an earlier letter to Mersenne of June 1638 (*Oeuvres*, Vol. 2, pp. 154–168). Proofs of Fermat's theorem were eventually found, by Cauchy in 1813 and Legendre in 1816.[124]

A brief scrap of 11 lines (*Oeuvres*, Vol. 10, pp. 297–298) merely gives a statement of Fermat's theorem and is obviously datable to 1638. A following scrap of 8 lines gives the proof of a theorem known to the Greeks, that 8 times a triangular number plus 1 is a square number. This is done in cossic characters, and hence the date of the piece must be much earlier than 1638. Rule 14 of the *Regulae*, on the use of illustrative figures, gives the drawing of a triangular number as one of the examples. The letter to Stampioen of 1633 containing the formula for pentagonal numbers has been mentioned already, n. 20. This letter has a bearing on the numerical data in the large table in Part II of the present manuscript. It discusses a problem involving four of the semiregular solids inscribed in four spheres tangent to each other, with each sphere tangent to a circumscribing fifth sphere.

Figurate numbers are now of only passing interest, and are mentioned merely to state that the Greeks had such things and give a few examples. Nevertheless the subject was quite popular in the past; Dickson's chapter on the subject cites over 200 names, including famous ones such as Pascal, Fermat, Euler, Legendre, and Cauchy. Various propositions in number theory appeared originally in a context of figurate numbers. Pyramidal numbers remain in problems relating to the number of shot in pyramidal piles, with triangular or square base, found in some textbooks on algebra.

Coxeter in 1974[125] used the term "polyhedral numbers" for the numbers of spheres close packed in the shape of five of the regular and semiregular solids. He refers to the lighthearted small work of Kepler, "The Six-Cornered Snowflake." Kepler treated the packing of spheres in triangular or square arrangements, and in space in pyramids with triangular or square base. In the latter two cases the spheres form the closest possible packing; each internal sphere is surrounded by and touches twelve others (this can be seen for the square-based pyramid by setting it on one of its triangular lateral faces). He also referred to rhomboids formed of close-packed spheres. This was in connection with speculations on the reason why snowflakes have six corners. Coxeter considered the tetrahedron, octahedron, truncated octahedron, and the cuboctahedron, composed of close-packed spheres, and developed general formulas for each. His formulas for the tetrahedron (obviously) and the octahedron (which can be formed from two square-based pyramids, one of side n and the other of side $n - 1$) are equivalent to those given by Descartes; Coxeter uses the length of a side minus one for his base in the formulas. The other three are different, as neither the Greek concept of polygonal numbers nor Descartes' concept of polyhedral numbers involved packing, but only arrangement, and only in a few special cases would there be close packing if the dots were spheres. Coxeter's type appears to be inapplicable to the other regular and semiregular bodies.

Notes

1. The full title of the *Méthode* as it appears on the title page is: Discours/ de la Methode/ Pour bien conduite sa raison, & chercher/ la verité dans les sciences./ Plus/ La Dioptrique./ Les Meteores./ et/ La Geometrie./ Qui sont des essais de cete Methode. It was published in Leyden in 1637 by Jan Maire and has 414 pages. See the Bibliography for a modern edition of the *Méthode* and one of the many English translations.

2. This statement is from the biography of René Descartes by Leslie J. Beck in *The New Encyclopaedia Britannica*, 15th ed., p. 599.

3. The account of the manuscripts is given by Baillet in *La Vie de Monsieur Des-Cartes*, Vol. 2, page 428. Details are recited, with quotations, in various places in the *Oeuvres de Descartes*, particularly Vol. 1, pp. xvi–xix, Vol. 10, pp. 1–14, 173–177, 207–212. The inventory, the contents of which had been published as early as 1656 by Borel (*Vitae Cartesii Compendium*, pp. 16–19), is given in *Oeuvres* Vol. 10, pp. 5–12.

4. Foucher de Careil, *Oeuvres inédites de Descartes*, Vol. 2, pp. 214–34.

Some description of Foucher de Careil's collection may be useful. His find was not an accident but a reward for persistent search. The account of the lost papers, and also the inventory were known and from several of Leibniz's letters it was known that he had had some Descartes manuscripts. Foucher de Careil's first search through the Leibniz papers in the library at Hanover was not successful, but on a second search he located a bundle of uninventoried Leibniz papers covered with the dust of ages (la poussière séculaire) in a neglected cupboard. Among these were a number of Leibniz copies of Descartes manuscripts; others were found among Leibniz papers in another compartment.

The first volume of Foucher de Careil's work starts with a 16-page Preface giving the history of the manuscripts and details of his search. Next comes an Introduction of 128 pages, which is an essay on the *Méthode*. The first manuscript in the book has the heading "Pensées de Descartes, annotées par Leibniz" and has 26 printed pages of Latin text with an equal number of facing pages of French translation. The manuscript carries the mention that the pages were commenced in January 1619, but the fourth page of the printed text refers in the past tense to an event which had occurred in 1620. The first 7 pages contain miscellaneous statements of a philosophical nature. Then follow comments on various topics in physics, algebra and geometry. A marginal note by Leibniz states that he discovered the manuscript and made a copy on June 1, 1676, and a note to one section indicates that it was copied June 3, 1676.

The second manuscript has the heading "Remarques de Descartes sur ses principes de philosophie"; it occupies 10 printed pages of Latin text and 10 pages of French translation. The third is "Observations Météorologiques et questions, tiré d'un manuscrit inédit de Descartes, in-4"; 14 and 14 pages. The date February 5, 1635 is on the fifth printed page, heading a shift in topic, and there is a note by Leibniz at the end. The fourth manuscript, concluding the first volume, is headed "Physiologie, d'après un manuscrit inédit de Descartes, que l'on conserve a Hanovre, copie de la main de Leibniz"; 28 and 28 pages. It has the appearance of seven short pieces, several having their own subheadings, joined together; the second part has the date November 1637, the third December 1637 and the fourth February 1648 (perhaps a misreading of 1638).

The second volume contains 24 letters of Descartes which Foucher de Careil had collected from various sources, and five additional manuscripts. The fifth (10 + 10 pages), sixth (24 + 24 pages) and seventh (38 + 38 pages) are under a general heading "Manuscrits Anatomiques." The fifth is dated 1631 and has notes by Leibniz, one of which states, "c'était un écrit de la jeunesse de Descartes," to which Foucher de Careil adds "Tout ce morceau en est la preuve." The eighth manuscript (4 pages) is medical and not translated. The ninth and last manuscript is the *De Solidorum Elementis*, which is not translated.

5. C.R. Acad. Sci., April 23, 1860 (see Bibliography). Prouhet was at that time editor of the *Bulletin de la Société Mathématique de France*.

6. *Revue de l'Instruction publique* of Nov. 1, 1860 (see Bibliography).

7. C. Mallet, [Review of Vol. 2 of Oeuvres inédites de Descartes], *Revue de l'Instruction publique* of Sept. 27, 1860 (see Bibliography).

8. Mallet, comments on Prouhet II, *ibid.*, Nov. 22, 1860, page 539; Prouhet, comments on preceding, *ibid.* Dec. 6, 1860, pages 571–572; Mallet, comments on preceding, *ibid.* Dec. 6, 1860, page 572.

9. Ernest de Jonquières, "Écrit posthume de Descartes", in *Mém. Acad. Sci.* for 1890 (see Bibliography). The translation and notes had already been printed in *Bibliotheca Mathematica*, 3F., 4 (1890) 43–55. This memoir was preceded by three notes in *Comptes rendus* (see Bibliography).

10. See Adam's account of the manuscript and history of the text, *Oeuvres* Vol. 10, pp. 257–263. See the Bibliography for a full description of this edition of the collected works.

11. Costabel appears to have no knowledge of the corrections made by the editors in Vol. 11, since his corrections include some (but by no means all) of theirs. He also adds some new considerations.

12. The identification marks on the two sheets of the manuscript are LH IV, 1, 4b Bl 1 on one and LH IV, 1, 4b Bl 15 on the other.

13. *Oeuvres* Vol. X (nouvelle édition), p. 687, n. on p. 265.

14. Three specimens of Descartes' handwriting which I happen to have on hand are on quarto sheets averaging 17×23 cm in size (with slight variations); these, if folded, would result in 11.5×17 cm octavo pages. Books of the period show that, in general, an octavo page would be about the same size.

15. [Evidently Leibniz was unable to read Descartes' manuscript clearly at this point. In the body of the table he wrote $\sqrt{\frac{11}{2}+\frac{1}{2}\sqrt{2}}$, in his annotation $\sqrt{\frac{17}{2}+6\sqrt{2}}$ remarking 'I know not why.' The *Oeuvres* text substitutes $1+\frac{3}{2}\sqrt{2}$ which is mathematically correct but far from the text. We must surely read $\sqrt{\frac{11}{2}+\frac{6}{2}\sqrt{2}}$, equivalent to the latter, but paleographically close to both expressions given by Leibniz. G.J.T.]

16. G. Milhaud, "L'oeuvre de Descartes," 86–89.

17. Milhaud obtained his information from Prouhet, who obtained his information from a 1799 history of mathematics by Kästner which contained a summary of Faulhaber's work. But Prouhet misinterpreted Kästner, and he and Milhaud give the impression that Descartes merely added some more polyhedra to the six given by Faulhaber, which is not the case. This matter is treated fully in Section 12, pp. 116-118, where the references are cited. Despite this mistake, it remains probably that Descartes listened to Faulhaber discoursing on his hobby.

18. Clavius, *Algebra, passim.*

19. See p. 7 above and Adam's account in *Oeuvres* Vol. X, pp. 259-262.

20. This period is well treated in general histories of mathematics. Those used most frequently in preparing this work are Kline, *Mathematical Thought from Ancient to Modern Times,* and Smith, *History of Mathematics.* Cajori in his *History of Mathematical Notations* gives numerous details throughout.

Viète's innovation of using letters for quantities was more than a shift in symbols. The consonants represented general classes of numbers and enabled equations to be written with general coefficients, whereas previously equations with specific numerical coefficients only were mainly if not solely considered. With Descartes' use of small letters, with x, y, z, for unknowns, and the indication of powers by arabic numeral exponents, we are almost at modern algebra; almost, as he did not use the modern equals sign.

It is evident from some of his writings that Descartes may not have shifted at first entirely from the old system to that of the *Géométrie;* the *Regulae ad Directionem Ingenii* show the use of capital letters to denote quantities in identities while at about the same time, according to the report by Beeckman, he was solving a quadratic equation in cossic symbols. Even in the case of Viète there was some development after his initial work of 1591, the *In artem analyticam isagoge.* In his *Ad Logisticen speciosam* (published posthumously in 1646), he was concerned with describing a new "species logistics," in contrast with number logistics which is calculation by numbers. For his new algebra he used capital letters to denote magnitudes, quantities, and developed the operations of addition, subtraction, multiplication and division, and ratios, proportions and identities, using letters. For example in the *Ad Logisticen speciosam* he gives the identity: "A squared, + A by B two times, + B squared, is equal to A + B squared." And in the geometrical examples he gives the sides and hypotenuse of a right-angled triangle as "Aq − Bq," "A by B two times," "Aq + Bq." (He used + for "plus," but = for "minus"). However, he still used the Latin word for "squared," or its abbreviation q, to denote the second power, etc. Equations with literal coefficients are not yet treated.

The date of Descartes' *Regulae ad Directionem Ingenii* is conjecturally placed in 1628 by the editors of the *Oeuvres* (Vol. 10, pp. 486–488), on the grounds that it could not have been written until Descartes abandoned the high life of Paris in 1628. But he did not leave Paris until December, when he went to a retreat somewhere in France, and he left France for Holland in March 1629. Thus a more plausible date for its composition is 1629. This date may well mark the beginning of the shift by Descartes to his new system. Rule 16 concerns the use of "highly abbreviated symbols." He states:

> Everything, therefore, which is to be looked upon as single from the point of view of the solution of our problem, will be represented by a single symbol which can be constructed in any way we please. But to make things easier we shall employ the characters a, b, c, etc. for expressing magnitudes already known, and A, B, C, etc. for symbolising those that are unknown. To these we shall often prefix the numerical symbols, 1, 2, 3, 4, etc., for the purpose of making clear their number,...

A power, $2a^3$, is represented, and the hypotenuse of a right-angled triangle with sides a and b is given as $\sqrt{a^2 + b^2}$. As is evident from the *Géométrie*, changes and additions came later. Perhaps these occurred by 1633. A letter to one Stampioen, dated by the editors of the *Oeuvres* as at the end of 1633, is published in Vol. 1, pp. 275–280. The original, which itself is a defective copy, has a marginal sentence which includes the general formula for pentagonal numbers (see Sec. 12 p. 119) in the form $(3xx - 1x)/2$. In this the cossic characters appear to have been abandoned, but the evidence is rather dubious.

21. Beck, "Descartes", p. 599.

22. Cf. Eneström's note on Descartes and cossic symbols in *Bibliotheca Mathematica* of 1905. This note presumably was the result of the consultation.

23. This remark was no doubt directed at Moritz Cantor, *Geschichte der Mathematik*, Vol. 2, p. 684, who states categorically that Descartes learned the cossic symbols from Faulhaber. Furthermore, while Faulhaber used the cossic symbols for the radix and the radix squared, for the cube he used 'Cub.'

24. Letter from Descartes to Beeckman, March 26, 1619, printed in *Oeuvres*, Vol. 10, pp. 154–160 (the relevant passage is on pp. 155–156).

25. Isaac Beeckman, "Algebrae Des Cartes specimen quoddam," printed in *Oeuvres* Vol. 10, pp. 333–335. This brief note by Beeckman of some of Descartes' ideas merits some study in its own right. In the *Géométrie* Descartes departed from the ancient and then current practice of considering a^2 as the surface of a square of side a and b^3 as the volume of a cube of side b; he said, "Here it must be observed that by a^2, b^3, and similar expressions, I ordinarily mean only the simple lines, which, however, I name squares, cubes, etc., so that I may make use of the terms employed in algebra." All powers were to be considered equally as lines, in effect as algebraic rather than geometric quantities. The note of 1628 expresses this idea in general terms with an illustration using 3 as the radix. There is a figure with five lines. The first is one unit in length. The second is 3 units in length and is marked with the

cossic symbol ℛ for the radix or quantity. The third is 9 units in length and is marked with the cossic symbol ⅔ for the quantity squared. The fourth line is 27 units in length and is marked with the cossic symbol ₡ for the quantity cubed. The fifth line is 81 units long, drawn zigzag, and is marked with the cossic symbol for the fourth power, which is ⅔⅔. A preceding figure shows what appears to be an intermediate stage for explanation. The note closes with the solution of a quadratic equation worked out in cossic symbols. See also Sec. 11, p. 95.

26. Snell, *Doctrinae triangulorum canonicae*, libri III, prop. VIII, pp. 120-122. Frajesi, "La teoria dell'uguaglianza dei triedri," pp. 224-227, states that the polar spherical triangle was described by the Persian Naser-Eddin (1201-1274) and that Viète in 1593 gave an obscure and confused description of a special case, as did several others.

27. Albert Girard, *Invention Nouvelle en l'Algebre*. This is a collection of three essays on arithmetic, algebra, and spherical trigonometry. The third, entitled "De la mesure de la superfice des triangles & polygones sphericques, nouvellement inventée," occupies the last 15 pages of the work (G1 verso to H4 verso, inclusive). It gives the areas of spherical triangles and polygons as well as the measure of solid angles. (See below p. 126, n.34).

A history of the topic is given by Vacca, "Notizie storiche sulla misura degli angoli solidi." He states that the area of spherical triangles was previously discovered by T. Harriot, as is shown by unpublished manuscripts.

28. The preceding discussion presupposes that the two parts of the manuscript were written at about the same time or, if written at different times, that the first part was done before the second. It is the order that is significant: if the first was written some time after the second, then some of the statements made would need modification. However, there is no evidence to indicate that the order of composition of the two parts was different from that in which they appear.

29. All references to Euclid, given by Book and proposition or definition number, are to Heath's translation.

30. tr. Morrow, pp. 300-301 (commentary on I 32). Heath repeats the proof of Proclus in his notes to Euclid I 32.

31. Proclus, tr. Morrow, pp. 301-302. Heath also gives Proclus' proof, and remarks that this property was not new with Proclus, since it was stated by Aristotle.

32. G. Pólya, *Induction and Analogy in Mathematics*, pp. 58,226.

33. Notes to Euclid XI Defs. 10 and 11, Vol. 3 pp. 265-268. Heron's definition (*Def.* 22, *Opera* IV pp. 28-30) is "A solid angle is in general the bringing together of a surface which has its concavity in one and the same direction to one point."

34. Girard, *Invention Nouvelle*, last essay (cf. n.27 above). In 1629 Girard gave the area of a spherical polygon as equal to the sum of its interior angles less the sum of the interior angles of a plane polygon having the same number of sides. Hence the term "spherical excess formula." He proved it for the

spherical triangle and then indicated that it could be proved for a general spherical polygon by dividing it into triangles. He also gave a corresponding expression for the measure of a solid angle; he uses the expression "inclination of the planes" for the dihedral angles. While he does indicate that the right angle unit of measure for angles can be used, he used degrees (since he used trigonometric functions), and gave numerical examples; he also introduced the concept of "degrez superficiels" (surface or areal degrees) for the spherical polygons and solid angles, the degrees for plane angles presumably being thought of as linear degrees. There is an interesting discussion of the units for angles.

35. See above, p. 36.

36. The text printed here (which is given, paragraph by paragraph, for ready comparison with the translation), is the one that I think should be read. Additions, deletions and changes from the reading of the manuscript are not noted as such in the text (for these the reader is referred to the critical edition above, pp. 22-29), but, where necessary, are explained in the comments.

The translation is intended to be a substantially literal rendering of the Latin text (rather than a literary one), as close to the original and as intelligible as possible. It was compared, when in first rough draft, with the French translation of Prouhet, allowing for the deficiencies in the text he had on hand, and with the Italian translation of Natucci, which were used to assist in revising the draft. It differs from these in various respects: for example, it is closer to the original in places, and avoids anachronistic terms. Furthermore, it is based on a more complete and more correct transcription of the handwritten Latin text. Explanatory additions are enclosed within square brackets. Supplements are in italics.

The comments vary in length and content. Most are explanatory elaborations. Proofs and derivations are given for various propositions; these are not necessarily those used by Descartes, who gave none, but are intended to show the truth of the propositions and are based on material in the manuscript itself with only a few exceptions. Most of the comments by Prouhet and Mallet were directed towards corrupt parts of Foucher de Careil's text and are not applicable to the correct text, but some which are still relevant are noted.

37. Prouhet and de Jonquières both keep this text and render it "plane angles," which they take as referring to the face angles. They attempt to remove the mathematical error by saying that Descartes is referring to the plane angles of the *exterior* solid angle.

38. G. Pólya, *Induction and Analogy in Mathematics.* He poses the proof of the proposition (given by a translation of Descartes' statement) on p. 57, and indicates his solution on p. 226.

39. Coxeter, *Regular Polytopes,* p. 24.

40. Descartes, *Regulae ad Directionem Ingenii* in *Oeuvres,* Vol. 10, p. 368. In Haldane and Ross's translation, p. 7.

41. Kepler, *Harmonice Mundi,* Book II at pp. 78 and 83.

42. This sentence and the following one (Paragraph 5) were joined as a single sentence in the Foucher de Careil printing, which also contained other

errors. Prouhet stated that it was unintelligible. Mallet separated the two and gave a significance which, except for statements based on incorrect words, is the same as that given here; Prouhet said that he doubted Mallet's interpretation but that he could not offer anything better.

43. The copying of such a trivially obvious proposition by Leibniz would suggest that he did not omit anything of substance.

44. Archimedes, ed. Heiberg, I p. 76; tr. Heath p. 24. Archimedes gives the definition of a 'solid rhombus,' *ibid.* Def. 6 (p. 6 Heiberg, p. 3 Heath).

45. Heath, *Euclid,* I def. 22, pp. 189-190, discusses the various meanings of rhombus and rhomboid.

46. Branko Grünbaum, personal communication.

47. Branko Grünbaum, *Convex Polytopes,* p. 287.

48. These are the polyhedra with 12 faces and in which all solid angles are trihedral, of which there are 7595; see Duijvestijn and Federico, "The number of polyhedral graphs."

49. De Jonquières states that a polyhedron is determined absolutely by the number of faces and their respective kinds. This is the case with the example, which is very simple, but is not true in general. For instance, if 7 faces and 24 plane angles are given, the number of solid angles will be 7 and the conditions are fulfilled by 7 topologically distinct polyhedra, including 5 different ones which have the identical combination of types of faces. These latter are Nos. 38 to 42 in the illustrated catalogue given in Federico, "Polyhedra with 4 to 8 faces."

50. Pappus, *Synagoge* V, 33ff. (ed. Hultsch, Vol. I p. 350ff; cf. *ibid.* Vol. III pp. 1138-1164). On the topic of 'isoperimetry' among the ancient Greeks, which goes back to Zenodorus (early 2nd century B.C.), see Heath, *A History of Greek Mathematics,* Vol. 2, pp. 206-213.

51. Prouhet stated that he could not give any sense to this sentence. Mallet could not either, and suggested that perhaps there was some omission or transposition and further remarked that the sentence appeared to be out of place. This last may very well be the case as it is at the very bottom of the sheet and may have been first omitted in the copying and then picked up when it was seen that there was still room at the bottom for another sentence.

52. This consisted, essentially, of the theorems concerning the five regular polyhedra found in Euclid XIII, and the description of the thirteen semiregular bodies, discovered by Archimedes, but known only from the account by Pappus in Bk. V of his *Synagoge* (see p. 101).

53. Panofsky, "Dürer as a Mathematician," p. 618.

54. This work existed only in manuscript until it was printed in 1916 by Mancini, "L'opera *De corporibus regularibus* di Pietro Franceschi." Mancini's thesis is that Pacioli copied from Franceschi.

55. Luca Pacioli, *Diuina Proportione.* This work was written in 1497. Less than a third of it is concerned with the Divine Proportion; the rest is devoted to solid bodies. Pacioli is said to have taken material freely from others: See Smith, *History of Mathematics,* pp. 252-254, who states with respect to the regular bodies, "Pacioli here takes his material freely from Franceschi's

work,...." Mancini (n. 54) is concerned with Pacioli's copying from Franceschi in the *Diuina Proportione*. Franceschi does not have perspective drawings but only various sections and while Pacioli does show perspective drawings of the same seven polyhedra discussed by Franceschi, he shows and discusses numerous solids not even mentioned by Franceschi.

A contemporary portrait of Pacioli shows him holding a solid dodecahedron in one hand. Evidently models were made during this period; the numerous perspective drawings in his work would have been extremely difficult if not impossible to make without models.

56. Albrecht Dürer, "Underweysung der messung mit dem zirckel und richtscheyt." This work is summarized by Panofsky, "Dürer as a Mathematician."

57. Wentzeln Jamitzer, *Perspectiva Corporum regularium*. Our Figs. 21, 22 and 23 are taken from this work.

58. Jacques Ozanam, *Dictionaire mathématique*, and Abraham Sharp, *Geometry Improv'd*, Part 2, "A Concise Treatise of Polyhedra or Solid Bodies of Many Bases."

59. To give just two modern examples, Coxeter, *Regular Polytopes*, p. 23, refers to it as "Descartes' Formula," and Buckminster Fuller, *Synergetics*, p. 54, also refers to Descartes in connection with it.

60. Bertrand, "Remarque à l'occasion de la Note précédente."

61. Poinsot, "Note sur la théorie des polyèdres."

62. That Descartes had any notion of topology is scouted by Milhaud, "L'oeuvre de Descartes," and by Lebesgue (see p. 78).

63. See e.g. Kreyszig, *Introduction to Differential Geometry and Riemannian Geometry*, pp. 209–214.

64. J.K. Becker, "Über Polyeder," and L. Lalanne, "Relations entre les quantités angulaires des polyèdres convexes."

65. L. Euler, *Elementa doctrinae solidorum* and *Demonstratio nonnullarum insignium proprietatum quibus solida hedris planis inclusa sunt praedita*.

66. Leonhard Euler, Letter of Nov. 3/14, 1750, to Christian Goldbach, Juškevič and Winter, *Leonhard Euler und Christian Goldbach, Briefwechsel 1729-1764*, pp. 332-333.

67. H. Freudenthal, "Leibniz und die Analysis situs," p. 616.

68. George Pólya, "Guessing and proving."

69. L. Euler, *Elementa doctrinae solidorum*, Opera 26, p. 73.

70. Imre Lakatos, *Proofs and Refutations*, p. 6.

71. E.g., Pólya, *Induction and Analogy in Mathematics*, pp. 35–43, 52–53, 223–224.

72. H. Lebesgue, "Remarques sur les deux premières démonstrations du théorème d'Euler rélatif aux polyèdres." See also p. 71.

73. L. Poinsot, "Mémoire sur les polygones et les polyèdres."

74. A.L. Cauchy, "Recherches sur les polyèdres"; L'Huilier (Lhuilier), "Démonstration immédéate d'un théorème fondamental d'Euler sur les poly-

èdres"; and Lhuilier, "Mémoire sur la polyédrométrie." These three papers are listed in the probably order in which they appeared, all in 1813. The third one (which refers to the first) is a summary and review of the earlier Lhuilier paper, by Gergonne, who added proofs and other material of his own.

75. Steiner, "Leichter Beweis eines stereometrischen Satzes von Euler" and "Anmerkungen zu den Aufsatze No. 18."

76. von Staudt, *Geometrie der Lage*. The proof of von Staudt is used by Coxeter, *Regular Polytopes*, p. 24.

77. J.B. Listing, "Der Census räumlicher Complexe"; A.F. Möbius, "Theorie der elementaren Verwandschaften"; C. Jordan, "Recherches sur les polyèdres."

78. Quoted by Pont (see p. 78).

79. May, "Historiographic vices," discusses "the fallacious attribution of results that are logical antecedents or consequences of established knowledge." He indicates that plausible inferences can be drawn "from directly documented knowledge, but valid argument for such inferences cannot rest on logical connections alone. It requires historical analysis of the knowledge, thinking, and environment of the persons involved."

80. Jacques Hadamard, *The Psychology of Invention in the Mathematical Field*, pp. 48–54.

81. Richard Baltzer, "Geschichte der Eulerschen Satzes von den Polyedern und der regularen Sternpolyeder."

82. Richard Baltzer, *Die Elemente der Mathematik*, Vol. 2, p. 207.

83. J.K. Becker, "Über Polyeder," p. 338.

84. De Jonquières [1].

85. De Jonquières [2].

86. De Jonquières [3].

87. W. Killing and H. Hovestadt, *Handbuch des Mathematischen Unterrichts*, pp. 267–272.

88. H. Lebesgue, "Remarques sur les deux premières démonstrations du théorème d'Euler rélatif aux polyèdres."

89. Jean-Claude Pont, *La Topologie Algébrique*, p. 13.

90. Cajori, *A History of Mathematical Notations*, Vol. 2, pp. 316–317.

91. E. Steinitz and H. Rademacher, *Vorlesungen über der Theorie der Polyeder*, p. 9.

92. Delachet, *La géométrie contemporaine*, p. 99. Fréchet and Fan (see n. 93).

93. Fréchet and Fan, *Introduction à la topologie combinatoire*, p. 25; English translation, p. 21.

94. D. Hilbert and S. Cohn-Vossen, *Anschauliche Geometrie*, pp. 254–280; English translation, pp. 290–295.

95. G. Pólya, *Induction and Analogy in Mathematics*, and G. Pólya, *Mathematical discovery*, Vol. II.

96. Pólya, *Induction and Analogy in Mathematics*, p. 56.

97. Jean-Claude Pont, *La Topologie Algébrique*, pp. 8–13.

98. Imre Lakatos, *Proofs and Refutations*, p. 6.

99. Nicomachus of Gerasa, *Introduction to Arithmetic*, (tr. D'Ooge) pp. 241–249; Théon de Smyrne, *Exposition des Connaissances Mathématiques Utiles pour la Lecture de Platon*, pp. 50–71.

100. Diophantus *On Polygonal Numbers*, in *Diophantus of Alexandria*, tr. T.L. Heath, pp. 247–259.

Descartes was familiar with Diophantus, as he states in the *Regulae ad Directionem Ingenii*. There were two Latin translations, the first by Xylander (Wilhelm Holtzman) published in 1575 in Basel, and the second by Bachet de Méziriac published in Paris in 1621 (the book whose margins were too small for Fermat). Tannery states that Descartes did not know Bachet's translation and had studied that of Xylander (*Oeuvres*, Vol. 10, p. 298). The context indicates that Tannery dated the manuscript to 1620 and hence concluded that Descartes did not know Bachet.

101. Iamblichus, *In Nicomachi Arithmeticam Introductionem*, pp. 58ff.

102. Heath, *A History of Greek Mathematics*, Vol. 2, pp. 213, 514–517; Dickson, *History of the Theory of Numbers*, Vol. 2, chap. 1, (Polygonal, Pyramidal and Figurate Numbers), pp. 1–3; Cohen and Drabkin, *A Source Book in Greek Science*, pp. 7–9.

103. The triangular numbers are commonly illustrated as balls, spherical shot or billiard balls, arranged in a triangular tray; thus, the group of 15 billiard balls racked up in the triangular rack is $p(3,4)$, 3 sides and 4 balls along each side. The balls are arranged in hexagonal packing; each interior ball touches each of its six neighbors, the closest horizontal packing possible. In the case of square numbers the dots are in a square lattice array and when represented by balls in a square tray each interior ball touches four others, a different kind of packing. But in the case of pentagonal and higher polygonal numbers as shown, there is neither of these types of packing. Packing is irrelevant to polygonal numbers.

104. Diophantus, tr. Heath, p. 252.

105. His figures illustrate this fact geometrically. In his series of pentagonal numbers each is shown by a square array with a triangle of side one less placed on its upper side. This is done by Theon as well, who also illustrates the series of hexagonal numbers by the series of pentagonal numbers with another triangle added to the opposite side of the square. This form could not very well have been used with higher polygons, but these do not appear to have been illustrated. The regular polygonal forms, as illustrated in Figs. 11 and 12, are required here, as they will be used as faces of polyhedra.

106. The description by Iamblichus of the formation of polygonal numbers (as translated in Cohen and Drabkin, p. 9) is as follows:

> In the representation of polygonal numbers two sides in all cases remain the same, and are produced; but the additional sides are included by the application of the gnomon and always change. There is one such additional side in the case of the triangle, two in the case of the square, three in the case of the pentagon, and so on indefinitely, the difference between the number of sides of the polygon and the number of sides which change being 2.

107. It is a modification of the figures given in Nicomachus (tr. D'Ooge), p. 244, and Cohen and Drabkin, p. 8.

108. The most common representations of pyramidal numbers are pyramidal piles of balls (shot, billiard balls, cannon balls) with a triangular or square base. These are packed in the closest form of packing, each interior ball touching twelve others. This is not the case with higher ones: the successive layers of polygons with the radix diminishing by one must be held in position to form a geometric pyramid as can be seen from Figs. 11 and 12. Thus packing is irrelevant to pyramidal numbers. (The reason for mentioning packing and its irrelevance will appear in Section 12 in connection with a work of Kepler of 1611 and a paper of Coxeter of 1974 which will be discussed there, pp. 119-120.

109. See Fine, *A College Algebra*, Chapter 23, "Arithmetical Progressions of Higher Orders," pp. 364-370, for a general treatment.

110. There appears to be room for a study and coordination of Descartes' views on units and measure. The subject appears in the *Géométrie* and is treated under Rules 15 and 18 of the *Regulae ad Directionem Ingenii*; the report by Beeckman of some of Descartes' algebra (referred to in Section 4, p. 32) is related to the statements on the subject in the *Géométrie*.

111. The paragraphs numbered here as 22 and 20 have been interchanged from their location in the manuscript, to avoid discussing the regular polyhedral numbers before the conclusion of the treatment of the polygonal numbers. Perhaps this is an instance of the shuffling of the original sheets after they were spread out to dry out from their immersion in the Seine. If the paragraphs here numbered 20, 21, 22 were each on a separate sheet, as could very well have been the case from their size, an interchange could have been possible.

112. As before (see p. 94), Descartes uses 'R' for 'radices' (the sides of the gnomon) and 'A' for 'anguli' (the vertices). In these tables he also uses 'F' for 'facies' (the faces). We have again prefixed a column giving n, the order of the number.

113. Pappus, ed. Hultsch V 33-36, II, pp. 350-358; tr. Commandino, pp. 83-84; tr. ver Eecke, pp. 272-277.

Descartes was familiar with Pappus, as he states in the *Regulae ad Directionem Ingenii* (*Oeuvres* X, p. 376), and parts of the *Géométrie* are taken up with the Problem of Pappus. It is evident that he derived his information concerning the Archimedean solids from Pappus and would have used the Commandino Latin text of 1588 or the 1589 (Venice) or the 1602 (Pesaro) reprint. We give here a summary of Pappus' account.

Pappus begins by referring to the perfect nature of the sphere and the fact that philosophers had stated that the sphere had the greatest volume of all the figures having the same surface area. But this had not been demonstrated. As in the preceding he had shown that the circle had the greatest area of all polygons having the same perimeter, he will now attempt to demonstrate that the sphere has the greatest volume of all the figures having the same surface area. It is possible to imagine a large number of solids with surfaces of various kinds but those which appear regular should rather be considered. These are not only the five figures of the divine Plato, but also "those which were discovered by Archimedes, thirteen in number, and are bounded by polygons which are equilateral and equiangular but not congruent."

Then follows a description of the thirteen solids: this is done simply by stating the total number of faces and the number of each type of face present in each solid.

The remaining (and largest) part of the passage is taken up with calculating the number of solid angles and the number of sides in each body. First, some general statements are made.

As to solid angles: if, for polyhedra whose solid angles are surrounded by three plane angles, one determines simply the number of plane angles of the faces of the polyhedron, the number of solid angles is the third of the number obtained; while for polyhedra whose solid angles are surrounded by four plane angles, the number of solid angles is one fourth of the number of plane angles. Likewise for polyhedra whose solid angles are surrounded by five plane angles, the number of solid angles is one-fifth the number of plane angles.

As to the number of sides: the "number [of sides of all the faces] is evidently equal to the number of plane angles; but since each side is common to two faces, it is evident that the number of sides of the polyhedron is the half of this number."

We give Pappus' first calculation *verbatim:*

> Consequently, since the first of the thirteen non-homogeneous polyhedra [the truncated tetrahedron, dealt with in this paragraph of Descartes] is bounded by 4 triangles and 4 hexagons, it has 12 solid angles and 18 sides; for the angles of the four triangles are 12 in number and the sides are 12 in number, while the angles of the four hexagons are 24 in number and the sides 24 in number; thus, the total number obtained being 36, the number of solid angles is necessarily the third of the number just given, since each solid angle of the polygon is surrounded by three plane angles, and the number of sides is half this number, that is to say 36; so that there are 18 sides.

The calculation is carried out for each of the thirteen polyhedra, although more briefly than the above for some, and only the results given for others.

After the above treatment of the thirteen solids, Pappus drops them "for the moment," because they are less regular and it is convenient to compare the sphere with the five regular figures, since these have all faces congruent and all solid angles composed of equal plane angles and hence are more regular than the others. The rest of Book V is concerned with isoperimetric problems relating to these, with no further mention of the Archimedean solids.

Pappus does not give names to the Archimedean solids. The term he employs for the face of a polyhedron is usually "base" ($\dot{\epsilon}\delta\rho\alpha$), but occasionally he uses "plane" ($\dot{\epsilon}\pi\dot{\iota}\pi\epsilon\delta o\nu$).

Several statements and ideas in Part I of the manuscript are obviously derived from Pappus. In Part II, too, the descriptions of Archimedean solids in the manuscript follow the order in Pappus, giving the number of faces and the number of each type, as well as the number of solid angles and sides, which were calculated by Pappus.

114. Kepler, *Harmonice Mundi,* Book II. Cf. pp. 60-61. As there stated, he discussed the regular and Archimedean solids, as well as others. He gave to the Archimedean solids the names which are still used today (and used by us). Some of them now have alternative names: the Rhombicuboctahedron is also known as the Small Rhombicuboctahedron, while the Truncated Cubocta-hedron is known as the Great Rhombicuboctahedron, and the Rhombicosi-dodecahedron and Truncated Icosidodecahedron are also known as the Small and Great Rhombicosidodecahedron, respectively.

There is no evidence in the manuscript that Descartes was aware of Kepler's work, published in 1619.

115. It is so translated by Hardie in his version of Kepler's "De nive sexangula."

116. Numerical values to six decimal places of the volumes of the regular and semiregular solids, with unit edge, are given by Berman, "Regular-faced Convex Polyhedra." These have been used to verify the expressions in this column; in the manuscript there are 3 errors, which have been noted and corrected on p. 28.

The title of Berman's paper refers to the general class of polyhedra in which all faces are regular polygons—regular-faced. These are divided into three classes. (1) The five regular solids, in which the faces are congruent regular polygons and the corners are congruent solid angles. (2a) The thirteen Archimedean solids, in which the faces are not all congruent but the corners are; and (2b) the two infinite series of prisms (excluding the cube which is included in (1)) and antiprisms (excluding the octahedron which is included in (1)); these, which have the same properties as the Archimedean solids, were first pointed out by Kepler. (3) The remaining solids, in which the corners are not congruent.

The third class has precisely 92 members; in all but two the faces as well as the corners are not congruent; two, the triangular and pentagonal bipyramids, have congruent equilateral triangles for faces. The complete set was first given (but not proven to be complete, which was done later by V.A. Zalgaller, see Bibliography) by Norman W. Johnson, "Convex polyhedra with regular faces." both Johnson and Berman present tables of the complete set of all three classes of regular-faced polyhedra with a variety of data relating to each, the two series being given in general terms. Berman gives the planar developments, and photographs of models, of each of them, 110 in number, plus a representative of each of the two series.

117. It was known to the ancient Greeks that each of the regular and Archimedean solids could be inscribed in a sphere. The radii of the circumscribing spheres of each are given by Robert Williams, *The Geometrical Foundation of Natural Structure,* pp. 63-67, 72-97 (a book which contains perspective drawings and a great deal of numerical information about each of the 18 solids). In 12 instances expressions similar to those in this column are

given and in the others only their numerical values, to four decimal places. The algebraic expressions for the radii of the circumscribing spheres for all 13 Archimedean solids are given by Papadatos, *Archimedes* (who also gives the expressions for the volumes). Only one error was found in the manuscript (corrected on p. 28).

118. It was known before Descartes that various of the Archimedean solids could be produced from the regular ones by truncating the corners in a uniform manner. A regular pyramidal piece is cut off each corner. If each cut goes less than half of the way (by an appropriate amount) into the edges incident to a corner, the results are the five Archimedean solids known as the truncated tetrahedron, cube, octahedron, dodecahedron and icosahedron. If the cut goes exactly halfway into each edge only two others are formed: the tetrahedron yields simply another tetrahedron; the cube results in the cuboctahedron as does the octahedron, hence the name; the dodecahedron results in the icosidodecahedron as does the icosahedron, hence the name. The cut can go more than halfway down each edge, thus overlapping, but this only produces four already obtained. All the above are produced from single truncations of the five regular solids. Four others are produced in other ways. Two of the thirteen, those not mentioned in the manuscript, are not producible by truncations.

Alan Holden, *Shapes, Space, and Symmetry*, pp. 40–41, has photographs of models showing the progressive truncations of the cube, octahedron, dodecahedron and icosahedron.

The first entry in the last column refers, not to the regular truncations, but to the fact that the tetrahedron can be produced from the cube by cutting off four corners in such a way that one always cuts along diagonals of the faces of the generating cube. The second line probably refers to the fact (noted by de Jonquières, "Écrit posthume de Descartes," p. 375) that by joining the midpoints of the sides of each triangular face of the tetrahedron (of side two units) and cutting off the four vertices along the lines so formed, one produces the octahedron (of side one unit). Note, however, that it can also be taken as referring to the first item in the table, since each of the pieces cut off in the above procedure is itself a tetrahedron (of side one unit). Perhaps the relationships of the other four regular bodies were in the original manuscript and omitted by Leibniz.

The ratio of the edge of the derived solid to the edge of the solid from which it can be derived is given for each of the eleven Archimedean solids by Cundy and Rollett, *Mathematical Models*, pp. 94–109. The ratios given by Descartes agree with those given in that book. However, the manuscript omits some derivations: the truncated octahedron (line 8) can also be derived from the cube; the truncated cube (line 9) can also be derived from the octahedron; the truncated icosahedron (line 11) can also be derived from the dodecahedron; and the truncated dodecahedron (line 13) can also be derived from the icosahedron.

119. Prouhet I. Kästner, *Geschichte der Mathematik*, Vol. 3, pp. 111-152.

120. I have examined (in photocopy) the copy of this work in the Niedersächsische Landesbibliothek at Hanover. It has written notes which may be in the hand of Leibniz on some of the pages.

121. The table (on a folding sheet) is headed "Inexaustae Scientiae Tabula secretissima Arithmetices Arcana pandens." The table is inexhaustible because it can be extended to infinity to the right and downward. It is in fact the Pascal triangle in the position in which Pascal showed it, that is, the ordinary form rotated 45° counterclockwise. A first row of ones is only indicated. The table contains 8 columns, the first being the natural numbers in order. The second column gives the successive summations of the numbers in the first column, that is, the triangular numbers. The third column gives the successive summations of the numbers in the second column, which are the triangular pyramidal numbers. The next column proceeds in the same way, forming the "corporeal" bodies of the second kind, corresponding to "hyperpyramids" with triangular base, and so on to the last column. Each number in the table is the sum of the number to its left and the number immediately above it. Reading from southwest to northeast you get the binomial coefficients (except the initial 1, as a lefthand column of ones is not present).

Descartes may have been referring to Faulhaber's table when he refers to the possibility of placing the results of some formulas in infinitely extending tables "imitating the German cabala" (*Oeuvres*, Vol. 10, p. 297).

The series in the table all have an ultimate common difference of 1. The text gives cossic equations of the fourth degree (the symbol for the fourth power of the radix is a double second power symbol) for series with an ultimate common difference of 2, 3 and 4, corresponding to square, pentagonal and hexagonal based pyramids.

Cantor refers to Faulhaber's work on arithmetical series, *Vorlesungen* Vol. 2, pp. 683-684. Papers concerning arithmetical series of higher orders (neither of which refers to Faulhaber or any other source) are Pollock's "On the Extension of Fermat's Theorem of the Polygonal Numbers" and Funkenbusch's "Hyperspacial figurate progressions."

122. Dickson, *History of the Theory of Numbers*, Vol. 2, p. 5, lists a dozen arithmetic texts of the 16th century which include polygonal numbers.

123. Dickson, Vol. 2, p. 16. I have not seen the book to which Dickson refers, F.W. Marpurg, *Anfangsgründe des progressionalcalculs*, p. 307; G.S. Klügel, *Mathematisches Wörterbuch*, Vol. 3, pp. 827-828.

124. Dickson, *History of the Theory of Numbers*, Vol. 2, p. 18.

125. Coxeter, "Polyhedral Numbers."

Bibliography

Archimedes: *Archimedis Opera omnia,* edited by J.L. Heiberg, Vol. 1, 2nd edition, Leipzig, 1910. [English translation:] T.L. Heath, *The Works of Archimedes,* edited in modern notation. Cambridge, 1897 [reprinted New York (Dover), n.d.]

Baillet, Adrien: *La Vie de Monsieur Des-Cartes,* 2 vols. Paris, 1691 [reprinted Hildesheim and New York, 1972].

Baltzer, Richard: *Die Elemente der Mathematik,* 2 vols. Leipzig, 1860, 1862.

Baltzer, Richard: Geschichte der Eulerschen Satzes von den Polyedern und der regularen Sternpolyeder, *Monatsbericht der Preussischen Akademie der Wissenschaften zu Berlin* **64,** 1861, pp. 1043-1086.

Beck, Leslie J.: Descartes, René, *The New Encyclopaedia Britannica,* 15th edition., Vol. 5, pp. 597-602. Macropaedia, 1975.

Becker, Joh. Karl: Über Polyeder, *Zeitschrift für Mathematik und Physik* **14,** 1869, pp. 65-76; Nachtrag zu dem Aufsatze über Polyeder, *Zeitschrift für Mathematik und Physik* **14,** 1869, pp. 337-343.

Beeckman, Isaac, Journal: *Journal tenu par Isaac Beeckman de 1604 à 1634,* publié... par C. de Waard, 4 vols. The Hague, 1939-1953.

Berman, Martin: Regular-faced Convex Polyhedra, *Journal of The Franklin Institute* **291,** 1979, pp. 329-352 plus seven pages of photographs.

Bertrand, J.: Remarque à l'occasion de la Note précédente [comment on Prouhet 1], *Comptes rendus des Séances de l'Académie des Sciences (Paris),* **50,** 23 avril 1860, pp. 781-782.

Borel, Pierre: *Vitae Renati Cartesii summi philosophi Compendium.* Paris, 1656.

Cajori, Florian: *A History of Mathematical Notations,* 2 vols. Chicago, 1928, 1929.

Cantor, Moritz: *Vorlesungen über Geschichte der Mathematik,* Zweiter Band, 2nd edition. Leipzig, 1900.

Cauchy, A.L.: Recherches sur les polyèdres, 1er Mémoire, *Journal de l'École Polytechnique* **9,** 1813, pp. 68-86 [reprinted in *Oeuvres complètes d'Augustin Cauchy,* II Série, Tome 1, pp. 1-25. Paris, 1905.].

Clavius, *Algebra: Algebra Christophori Clavii e Societate Iesv.* Rome, 1608 [reprinted in *Christophori Clavii Bambergensis e Societate Iesv Opervm Mathematicorvm Tomus Secundus.* Mainz, 1611 (the third item in the book, separately paged, and dated 1612)].

Cohen, Morris R., and Drabkin, I.E.: *A Source Book in Greek Science.* Cambridge, Mass., 1958.

Coxeter, H.S.M.: Polyhedral Numbers, in *For Dirk Struik: Historical and Political Essays in Honor of Dirk J. Struik*, edited by R.S. Cohen, J.J. Stachel, and M.W. Wartofsky, pp. 25–35. Dordrecht, 1974.

Coxeter, H.S.M.: *Regular Polytopes*, 2nd edition. New York, 1963.

Cundy, H. Martyn, and Rollett, A.P.: *Mathematical Models*. Oxford, 1952.

Delachet, A.: *La géométrie contemporaine*. Paris, 1957.

Descartes, *De Solidorum Elementis:* in *Oeuvres*, Vol. 10, pp. 265–276. [*Additions* in *Oeuvres*, Vol. 11, pp. 690–692.]

Descartes, *Géométrie: The geometry of René Descartes*, translated from the French and Latin by David Eugene Smith and Marcia Latham; with a facsimile of the first edition, 1637. Chicago and London, 1925 [reprinted New York (Dover), 1954].

Descartes, *Méthode:* Discours de la Methode, in *Oeuvres*, Vol. 6, pp. 1–515. [Partial English translation: E.S. Haldane and G.R.T. Ross, *The Philosophical Works of Descartes*, Vol. 1, pp. 81–130. Cambridge, 1911 (does not include the "Géométrie" and the other "Essays on this Method")].

Descartes, *Oeuvres: Oeuvres de Descartes*, publiées par Charles Adam & Paul Tannery. 12 vols. Paris, 1897–1913. Nouvelle présentation, en co-édition avec le Centre Nationale de la Recherche Scientifique. Paris, 1964–1975 [facsimile reprint of the original, with additional notes].

Descartes, *Regulae ad Directionem Ingenii:* in *Oeuvres*, Vol. 10, pp. 359–469. [English translation: E.S. Haldane and G.R.T. Ross, *The Philosophical Works of Descartes*, Vol. 1, pp. 1–77, Cambridge, 1911].

Descartes. *See also* Foucher de Careil; de Jonquières; Natucci.

Dickson, Leonard E.: *History of the Theory of Numbers*. 3 vols. (Carnegie Institution Publication No. 256). Washington, D.C., 1919.

Diophantus: Sir Thomas L. Heath [tr.], *Diophantus of Alexandria*, 2nd edition. Cambridge, 1910.

Dürer, Albrecht: *Underweysung der messung mit dem zirckel uñ richtscheyt in Linien ebnen unnd gantzen corporen*. Nürnberg, 1525. [Facsimile reprint, Zürich, 1966.]

Duijvestijn, A.J.W., and Federico, P.J.: The number of polyhedral graphs, *Mathematics of Computation* **87**, 1981, pp. 523–532.

Eneström, G.: [Note on Cantor's *Vorlesungen über Geschichte der Mathematik*, Vol. 2, 2nd edition, pp. 793–794]. *Bibliotheca Mathematica*, 3 Folge, **6**, 1906, pp. 405–406.

Euclid: *The Thirteen Books of the Elements*, Translated with Introduction and Commentary by Sir Thomas L. Heath, 2nd edition, 3 vols. Cambridge, 1926 [reprinted New York (Dover), 1956].

Euler, Leonhard: Demonstratio nonnullarum insignium proprietatum quibus solida hedris planis inclusa sunt praedita, *Novi commentarii academiae scientiarum Petropolitanae* **4** (1752/3), 1758, pp. 140–160 [reprinted in Euler, *Opera Mathematica*, Vol. 26, pp. 94–108].

Euler, Leonhard: Elementa doctrinae solidorum, *Novi commentarii academiae scientiarum Petropolitanae* **4** (1752/3), 1758, pp. 109–140 [reprinted in Euler, *Opera Mathematica*, Vol. 26, pp. 71–93].

Euler, *Opera: Leonhardi Euleri Opera Omnia*, Series Prima, *Opera Mathematica*, Vol. XXVI. Commentationes Geometricae, edited by Andreas Speiser. Zürich, 1953.

Euler: *See also* Juškevič and Winter.

Faulhaber, Johann: *Numerus figuratus; siue, Arithmetica analytica arte mirabili jnavdita nova constans.* [Frankfurt], 1614.

Federico, P.J.: Polyhedra with 4 to 8 faces, *Geometria Dedicata* 3, 1975, pp. 469–481.

Fine, Henry B.: *A College Algebra.* New York, 1901.

Foucher de Careil: *Oeuvres inédites de Descartes, précédées d'une introduction sur la méthode,* par M. le Cte. Foucher de Careil, 2 vols. Paris, 1859, 1860.

Frajesi, Attilio: La teoria dell'uguaglianza dei triedri nel suo sviluppo storico, *Periodico di Matematiche* 14, 1934, pp. 211–234.

Fréchet, Maurice, and Fan, Ky: *Introduction à la topologie combinatoire.* Paris, 1946. [English translation with notes by Howard W. Eves, *Initiation to Combinatorial Topology.* Boston, 1967].

Freudenthal, Hans: Leibniz und die Analysis situs, in *Homenaje a Millás-Vallicrosa,* Vol. I, pp. 611–621. Barcelona, 1954.

Fuller, R. Buckminster: *Synergetics: Explorations in the Geometry of Thinking* (in collaboration with E.J. Applewhite). New York, 1975.

Funkenbusch, William: Hyperspacial figurate progressions, *American Mathematical Monthly* 52, 1945, pp. 571–591.

Girard, Albert: *Invention Nouvelle en l'Algebre.* Amsterdam, 1629 [reprinted by D. Bierens de Haan, Leiden, 1884].

Grünbaum, Branko: *Convex Polytopes.* London and New York, 1967.

Hadamard, Jacques: *The Psychology of Invention in the Mathematical Field.* Princeton, 1945.

Heath, T.L.: *A History of Greek Mathematics.* 2 vols. Oxford, 1921.

Heath, T.L. *See also* Archimedes; Euclid.

Heron: *Heronis Alexandrini Opera quae supersunt omnia,* edited by J.L. Heiberg, Vol. IV. Leipzig, 1912.

Hilbert, D., and Cohn-Vossen, S.: *Anschauliche Geometrie.* Berlin, 1932. [English translation by P. Nemenyi: *Geometry and the Imagination.* New York, 1952].

Holden, Alan: *Shapes, Space and Symmetry.* New York and London, 1971.

Iamblichus: *Iamblichi in Nicomachi Arithmeticam Introductionem Liber,* edited by H. Pistelli. Leipzig, 1894.

Jamitzer, Wentzeln: *Perspectiva Corporum regularium.* Nürnberg, 1568. [Facsmile reproduction, Graz (Akademische Druck- und Verlagsanstalt), 1973].

Johnson, Norman W.: Convex polyhedra with regular faces, *Canadian Journal of Mathematics* 18, 1966, pp. 160–200.

de Jonquières [1]: E. de Jonquières, Note sur un Mémoire de Descartes longtemps inédit et sur les titres de son auteur à la priorité d'une découverte dans la théorie des polyèdres, *Comptes rendus des Séances de l'Académie des Sciences* 110, 1890, pp. 261–266.

de Jonquières [2]: E. de Jonquières, Écrit posthume de Descartes sur les polyèdres, *Comptes rendus des Séances de l'Académie des Sciences* 110, 1890, pp. 315–317.

de Jonquières [3]: E. de Jonquières, Note sur un Mémoire présenté, qui contient, avec le texte complet et revu de l'écrit posthume de Descartes: De solidorum elementis, la traduction et le commentaire de cet Ouvrage, *Comptes rendus des Séances de l'Académie des Sciences* **110**, 1890, pp. 677–680.

de Jonquières, Ernest: Écrit posthume de Descartes. *De Solidorum elementis.* Texte latin (original et revu) suivi d'une traduction française avec notes, *Mémoires de l'Académie des Sciences de l'Institut de France*, 2 Série, **45**, 1890, pp. 325–379. [Also printed separately, Paris, 1890, 55pp.]

Jordan, Camille: Recherches sur les polyèdres, *Comptes rendus des Séances de l'Académie des Sciences* **62**, 1866, pp. 1339–1341.

Juškevič, A.P. and Winter, E. (Eds.): Leonhard Euler and Christian Goldbach, Briefwechsel 1729–1764, *Abhandlungen der Deutschen Akademie der Wissenschaften zu Berlin*, Kl. f. Philosophie, 1965.1. Berlin, 1965.

Kästner, Abraham Gotthelf: *Geschichte der Mathematik*, 4 vols. Göttingen, 1796–1800.

Kepler, *Harmonice Mundi: Ioannis Keppleri Harmonices Mvndi Libri V.* Linz, 1619 [reprinted, Johannes Kepler, *Gesammelte Werke Band VI*, edited by Max Caspar. München, 1940; English translation of Book II in J.V. Field, Kepler's star polyhedra, *Vistas in Astronomy* **23**, 1979, pp. 109–141].

Kepler, Johannes: *The Six-Cornered Snowflake* [edited and translated by Colin Hardie]. Oxford, 1966.

Killing, W. and Hovestadt, H.: *Handbuch des Mathematischen Unterrichts.* Leipzig, 1913.

Kline, Morris: *Mathematical Thought from Ancient to Modern Times.* Oxford, 1972.

Klügel, Georg Simon: *Mathematisches Wörterbuch.* 5 vols. Leipzig, 1803–1831.

Kreyszig, Erwin: *Introduction to Differential Geometry and Riemannian Geometry.* Toronto, 1968.

Lakatos, Imre: *Proofs and Refutations: The Logic of Mathematical Discovery.* Cambridge, 1976.

Lalanne, L.: Relations entre les quantités angulaires des polyèdres convexes, *Comptes rendus des Séances de l'Académie des Sciences* **74**, 1872, pp. 602–603.

Lebesgue, H.: Remarques sur les deux premières démonstrations du théorème d'Euler, rélatif aux polyèdres, *Bulletin de la Société mathématique de France* **52**, 1924, pp. 315–336 [reprinted in Lebesgue, *Oeuvres*, Vol. 4 pp. 211–232].

Lebesgue, Henri Léon: *Oeuvres scientifiques*, 5 vols. Genève, 1972–1973.

Legendre, A.-M.: *Éléments de géométrie.* Paris, 1794 [12th edition Paris, 1823].

Lhuilier [S.A.J.]: Mémoire sur la polyédrométrie; contenant une démonstration directe du Théorème d'Euler sur les polyèdres, et un examen des diverses exceptions auxquelles ce théorème est assujetti, [Extrait] Par M. Gergonne, *Annales de Mathématique pures et appliquees*...rédigé Par J.-D. Gergonne **3**, 1812–1813, pp. 169–191.

L'Huilier, S.: Démonstration imédéate d'un théorème fondamental d'Euler sur les polyèdres, et exceptions dont ce théorème est susceptible, *Mémoires de l'Académie Impériale des Sciences de Saint-Pétersbourg* **4**, 1811, pp. 271–301.

Listing, J.B.: Der Census räumlicher Complexe oder Verallgemeinerung des Euler'schen Satzes von den Polyedern, *Abhandlungen der königlichen Gesellschaft der Wissenschaften zu Göttingen* **10**, 1862, pp. 97–180.

Mallet, C.: Compte rendu de Foucher de Careil (Ed.) *Oeuvres inédites de Descartes*, Vol. 2. *Revue de l'Instruction publique*, 27 sept. 1860, pp. 407–410.

Mancini, Girolamo: L'opera "De corporibus regularibus" di Pietro Franceschi detto della Francesca, usurpata da fra Luca Pacioli, *Atti della R. Accademia dei Lincei, Memorie della Classe di scienze morali, storiche e filologiche*, Roma, 1909, ser. 5 Vol. XIV, pp. 441–580.

Marpurg, Friedrich Wilhelm: *Anfangsgründe des progressionalcalculs überhaupt... und der construction der eckigten geometrischen körper*. Berlin and Stralsund, 1774.

May, Kenneth O.: Historiographic vices: I Logical attributions, *Historia Mathematica* **2**, 1975, pp. 185–187.

Milhaud, G.: L'oeuvre de Descartes pendant l'hiver 1619–1620, *Scientia* **23**, 1918, pp. 1–18, 77–90.

Möbius, A.F.: Theorie der elementaren Verwandschaften, *Berichte über die Verhandlungen der Königlich Sächsischen Gesellschaft der Wissenschaften, Math.-Phys. Kl.* **15**, 1863, pp. 31–68.

Natucci, A.: Il 'De Solidorum Elementis' di Cartesio, *Mathesis* **12**, 1920, pp. 117–127.

Nicomachus of Gerasa: *Introduction to Arithmetic*, translated by Martin Luther D'Ooge (University of Michigan Studies, Humanistic Series, Vol. XVI). New York, 1926.

Ozanam, Jacques: *Dictionaire mathématique*. Amsterdam, 1691 [also Paris, 1691].

Pacioli, Luca: *Diuina Proportione*. Milan, 1507 [Reprinted Milan (Fontes Ambrosiani), 1956].

Panofsky, Erwin: Dürer as a Mathematician, in *The World of Mathematics*, edited by James R. Newman, Vol. 1, pp. 603–621. New York, 1956.

Papadatos, Ioannes, Sp.: ΑΡΧΙΜΗΔΗ, τὰ 13 ἡμικανονικὰ Πολύεδρα [*Archimedes, The 13 semi-regular Polyhedra*]. Athens, 1978.

Pappus, Synagoge: *Pappi Alexandrini Collectionis quae supersunt*, edited by F. Hultsch, 3 vols. Berlin, 1875–1878. [Latin translation by Commandino: *Pappi Alexandrini Mathematicae collectiones a Federico Commandino Vrbinate in latinum conversae*. Pesaro, 1588.] [French translation with notes: *Pappus d'Alexandrie, La Collection Mathématique*, translated by Paul ver Eecke, 2 vols. Paris and Bruges, 1933.

Poinsot, L.: Mémoire sur les polygones et les polyèdres, *Journal de l'École Polytechnique* **4**, 1810, pp. 16–48.

Poinsot, L.: Note sur la théorie des polyèdres, *Comptes rendus des Séances de l'Académie des Sciences* **46**, 1858, pp. 65–79.

Pollock, Frederick: On the extension of Fermat's Theorem of the polygonal numbers to the orders of series whose ultimate differences are constant, *Proceedings of the Royal Society of London* **5**, 1843–1850, pp. 922–924.

Pólya, George: Guessing and proving, *California Mathematics* **1**, 1976, pp. 1–8.

Pólya, G.: *Induction and Analogy in Mathematics* [Vol. 1 of] *Mathematics and Plausible Reasoning*. Princeton, 1954.

Pólya, G.: *Mathematical discovery; on understanding, learning and teaching problem solving*, 2 vols. New York, 1962, 1965.

Pont, Jean-Claude: *La Topologie Algébrique des origines à Poincaré*. Paris, 1974.

Proclus: *A Commentary on the First Book of Euclid's Elements*, translated with introduction and notes by Glenn R. Morrow. Princeton, 1970.

Prouhet I: E. Prouhet, Remarques sur un passage des oeuvres inédites de Descartes, *Comptes rendus des Séances de l'Académie des Sciences,* **50,** 23 avril 1860, pp. 779-781.

Prouhet II: E. Prouhet, Notice sur la partie mathématique des *Oeuvres inédites de Descartes. Revue de l'Instruction publique,* 1er novembre 1860, pp. 484-487.

Ritter. F.: *See* Viète, Francois.

Sharp, Abraham: *Geometry Improv'd.* London, 1717.

Smith, David Eugene: *History of Mathematics,* 2 vols. Boston and New York, 1923, 1925 [reprinted New York (Dover), n.d.]

Snell [Snellius], Willebrord: *Doctrinae triangvlorum canonicae libri quatvor.* Leiden, 1627.

von Staudt, C.: *Geometrie der Lage.* Nürnberg, 1847.

Steiner, Jacob: Anmerkungen zu dem Aufsatze No. 18, *Journal für die reine und angewandte Mathematik,* Herausgegeben von A.L. Crelle, **3,** 1828, pp. 205-206 [reprinted in *Gesammelte Werke* 1, pp. 171-172].

Steiner, Jacob: Leichter Beweis eines stereometrischen Satzes von Euler, nebst einem Zusatze zu Satz X. S. 48 in 1 Heft dieses Journals, *Journal für die reine und angewandte Mathematik,* Herausgegeben von A.L. Crelle, **1,** 1826, pp. 364-367 [reprinted in *Gesammelte Werke* **1,** pp. 95-100].

Steiner, Jacob: *Jacob Steiner's Gesammelte Werke,* herausgegeben von K. Weierstrass, 2 vols. Berlin, 1881, 1882.

Steinitz, E. and Rademacher, H.: *Vorlesungen über die Theorie der Polyeder.* Berlin, 1934.

Struik, D.J. (Ed.): *A Source Book in Mathematics, 1200-1800.* Cambridge, Mass., 1969.

Theon of Smyrna: *Théon de Smyrne, Exposition des Connaissances Mathématiques Utiles pour la Lecture de Platon,* edited and translated by J. Dupuis. Paris, 1892.

Vacca, G.: Notizie storiche sulla misura degli angoli solidi e dei poligoni sferici, *Bibliotheca Mathematica,* 3 Folge, **3,** 1902, pp. 111-117.

Viète, François: Ad Logisticen Speciosam Notae priores, in *Francisci Vietae Opera Mathematica...recognita Opera Francisci à Schooten.* Leiden, 1646 [*reprinted* Hildesheim, 1970], pp. 13-41. [*French translation:* Première série de notes sur la logistique spécieuse par François Viète traduit par F. Ritter, *Bullettino di Bibliografia e di Storia delle Scienze Matematiche e Fisiche* **1,** 1868, pp. 245-276.]

Viète, François: *In Artem Analyticam Isagoge.* Tours, 1591. [Reprinted in *Francisci Vietae Opera Mathematica...recognita Opera Francisci à Schooten.* Leiden, 1646, pp. 1-12]. [French translation: Introduction à l'art analytique par Francois Viète traduit par F. Ritter, *Bullettino di Bibliografia e di Storia delle Scienze Matematiche e Fisiche* 1, 1868, pp. 223-244].

Williams, Robert: *The Geometrical Foundation of Natural Structure: A Source Book of Design.* New York, 1979.

Zalgaller, V.A.: Convex polyhedra with regular faces. *Seminars in Mathematics V.A. Steklov Mathematical Institute Leningrad* **2,** English translation, Consultants Bureau, New York, 1969.

Index of Persons